T0185718

SpringerBriefs in Applied Sciences and Technology

SpringerBriefs present concise summaries of cutting-edge research and practical applications across a wide spectrum of fields. Featuring compact volumes of 50 to 125 pages, the series covers a range of content from professional to academic.

Typical publications can be:

- A timely report of state-of-the art methods
- An introduction to or a manual for the application of mathematical or computer techniques
- A bridge between new research results, as published in journal articles
- A snapshot of a hot or emerging topic
- An in-depth case study
- A presentation of core concepts that students must understand in order to make independent contributions

SpringerBriefs are characterized by fast, global electronic dissemination, standard publishing contracts, standardized manuscript preparation and formatting guidelines, and expedited production schedules.

On the one hand, **SpringerBriefs in Applied Sciences and Technology** are devoted to the publication of fundamentals and applications within the different classical engineering disciplines as well as in interdisciplinary fields that recently emerged between these areas. On the other hand, as the boundary separating fundamental research and applied technology is more and more dissolving, this series is particularly open to trans-disciplinary topics between fundamental science and engineering.

Indexed by EI-Compendex, SCOPUS and Springerlink.

More information about this series at http://www.springer.com/series/8884

Anna Igual Munoz · Nuria Espallargas ·
Stefano Mischler

Tribocorrosion

 Springer

Anna Igual Munoz
Tribology and Interfacial Chemsitry Group
Ecole Polytechnique Fédérale de Lausanne
Lausanne, Switzerland

Stefano Mischler
Tribology and Interfacial Chemsitry Group
Ecole Polytechnique Fédérale de Lausanne
Lausanne, Switzerland

Nuria Espallargas
Tribology Lab, Department of Mechanical
and Industrial Engineering
Norwegian University of Science
and Technology
Trondheim, Norway

ISSN 2191-530X   ISSN 2191-5318   (electronic)
SpringerBriefs in Applied Sciences and Technology
ISBN 978-3-030-48106-3   ISBN 978-3-030-48107-0   (eBook)
https://doi.org/10.1007/978-3-030-48107-0

This Springer imprint is published by the registered company Springer Nature Switzerland AG
The registered company address is: Gewerbestrasse 11, 6330 Cham, Switzerland

Preface

Tribocorrosion can be considered a young research area which has experienced a boosting interest and development in the last 20 years mainly due to the increasing demand from engineering systems which are subjected to the combined action of wear and corrosion (i.e. biomedical implants, mining equipment, manufacturing, nuclear power plants, food processing devices and marine infrastructures). Tribocorrosion occurs in tribological contacts operating in corrosive environments and results in a material degradation or transformation caused by the interaction of chemical and tribological phenomena. This complex situation cannot be approached by simply considering both wear and corrosion separately. Thus specific tribocorrosion concepts and tailored experiments must be employed. This book is intended to constitute a toolbox for identifying and addressing tribocorrosion situations from an engineering point of view. It describes specific tribocorrosion concepts, models and experimental techniques as well as their application to practical situations in which mechanical and chemical phenomena act simultaneously. To do that, the book has been structured in eight chapters allowing for introducing the importance of the tribocorrosion phenomena (Chap. 1). Chapters 2 and 3 describe the basic corrosion and tribological concepts used in tribocorrosion. Chapter 4 establishes the tribocorrosion discipline through the identification of the main involved mechanisms (mechanical wear, corrosion and wear accelerated corrosion) and the involved variables. A practical way to approach tribocorrosion system is given in Chap. 5. Experimental techniques specifically designed to face a tribocorrosion problem are described in Chaps. 6 and 7. The last chapter of this book, Chap. 8, aims to give some practical examples of how to use tribocorrosion concepts in engineering systems.

This book addresses researchers, engineers and professionals from industry and academia dealing with engineering systems in which wear and corrosion takes place simultaneously, thus confronted with tribocorrosion situations. It constitutes an introductory book to those who want to move the first steps in the tribocorrosion field.

Lausanne, Switzerland   Anna Igual Munoz
Trondheim, Norway   Nuria Espallargas
Lausanne, Switzerland   Stefano Mischler

Contents

Symbols

A_a	Anode surface area, m^2
A_n	Nominal area of contact, m^2
A_r	Real area of contact, m^2
a	Radius of the contact area, m
a_c	Tafel coefficient
a_i	Area of the junctions, m^2
b_c	Tafel coefficient
C	Electrochemical corrosion rate, Kg/s
d_r	Thickness of reacted material, m
D_x	Amplitude displacement, m
E	Young's modulus, GPa
E'	Reduced Young's modulus, GPa
E_c	Potential of the cathode, V
E_{corr}	Corrosion potential, V
F	Faraday's number, 96485 C/mol
F_{eff}	Effective normal force, N
F_n	Normal load, N
F_t	Friction force, N
f_i	Load carried out by the junctions, kg
H	Surface micro-hardness, HV
h_0	Minimum fluid film thickness, mm
I	Current, A
I_a	Current of the anode, A
i	Current density, A/m^2
i_a	Current density of the anode, A/m^2
K	Wear coefficient
k	Constant
k_0	Proportionality constant
k_a	Constant
k_{chem}	Proportionality factor for chemical wear

k_{mech}	Proportionality factor for mechanical wear
L	Sliding distance, m
M_r	Atomic mass of the metal, g/mol
m	Mass, kg
m_r	Mass loss per unit surface and time, g/m^2 s
n	Stoichiometric coefficient of electrons
n_m	Number of moles reacted per unit surface and time, mol/m^2 s
OCP	Open circuit potential, V
P	Pressure, Pa
$P_{average}$	Average contact pressure, Pa
P_{max}	Maximum contact pressure, Pa
Q	Charge, C
Q_p	Passivation charge density, C/m^2
R	Radius of curvature, m
RR	Removal rate, m/s
R'	Reduced radius of curvature, m
R_a	Roughness average, μm
R_{dep}	Depassivation rate, m^2/s
R_q	Root mean square roughness, μm
S	Synergistic effect of the combined effect of wear and corrosion
SHE	Standard hydrogen electrode
T	Tribocorrosion rate, kg/s
TML	Total material loss, kg
t	Time, s
$t_{dry\ duty\ time}$	Time during which wear occurs, s
$t_{wet\ duty\ time}$	Time during the combined effect of corrosion and wear occurs, s
t_{static}	Time during which corrosion occurs (in implants), s
$t_{dynamic}$	Time during which the combined effect of corrosion and wear occurs (in implants), s
$t_{wet\ exposure}$	Time during which corrosion occurs (in excavator), s
u	Entraining velocity, m/s
V_{chem}	Chemical wear volume, mm^3
V_{corr}	Corrosion volume, mm^3
V_{mech}	Mechanical wear volume, mm^3
V_{tot}	Wear volume, mm^3
v_{corr}	Corrosion rate density, m/s
v'_{corr}	Number of metal moles dissolved into the solution per unit time, mol/s
v_s	Sliding velocity, m/s
v_{wear}	Wear rate, mm^3 /mN
W	Mechanical wear rate, Kg/s
σ_Y	Yield strength, Pa
λ	Lambda ratio
η	Viscosity of the solution, Pa s^{-1}
σ	Roughness

ρ	Density, kg/m^3
μ	Coefficient of friction
τ	Shear stress, Pa
ν	Poisson's ratio

Chapter 1
Tribocorrosion: Definitions and Relevance

In simple words, tribocorrosion is a material degradation resulting from the combined action of corrosion and wear. More in detail, the ASTM G40-10b defines tribocorrosion as a form of solid surface alteration that involves the joint action of relatively moving mechanical contact and a chemical reaction in which the result maybe different in effect than either process acting separately. This definition encompasses different types of interactions between corrosion and wear.

To better illustrate the tribocorrosion phenomenon, several practical examples are shown in Fig. 1.1:

1. The first case represents a pipe transporting slurry (Fig. 1.1a). The metallic components of the pipe will be always under the simultaneous effect of wear and corrosion because these hydraulic systems are designed to continuously transport liquids, except when they will be empty. The total material loss (TML) of these kinds of pipes is only due to the tribocorrosion damage (T) occurring during the whole duty time of the pipe ($t_{\text{wet duty time}}$).

$$TML = T \cdot t_{\text{wet duty time}} \tag{1.1}$$

2. In the case of biomedical implants such as hip joints replacing natural joints (Fig. 1.1b), materials are constantly in contact with human fluids, thus suffering from corrosion during their lifetime. Furthermore, some human activities (i.e. walking, running, jumping) also subject the joints to mechanical loading, thus this simultaneous action of wear and corrosion (tribocorrosion) occurs. In these cases, the total material loss (TML) from the artificial joint can be expressed as the sum of the corrosion effect (C) during static conditions (t_{static}) and the wear and corrosion acting simultaneously (T) during the periods of dynamic activities (t_{dynamic}).

$$TML = C \cdot t_{\text{static}} + T \cdot t_{\text{dynamic}} \tag{1.2}$$

© The Author(s), under exclusive license to Springer Nature Switzerland AG 2020
A. Igual Munoz et al., *Tribocorrosion*, SpringerBriefs in Applied
Sciences and Technology, https://doi.org/10.1007/978-3-030-48107-0_1

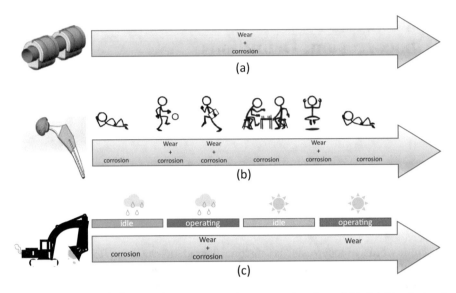

Fig. 1.1 Examples of tribocorrosion systems, **a** pipe transporting slurry, **b** hip joint implant and **c** excavator

3. The third case (Fig. 1.1c) corresponds to an excavator that is operating in an open environment (i.e. open air mine). Water from rain or wet ground generates corrosion of the excavating parts in idle conditions. The bucket teeth digging into the dry ground experiences wear due to abrasion by hard rocks and/or sand (dry wear rate W). When operating during rainy days, both wear and corrosion take place simultaneously, thus a tribocorrosion situation takes place (Tribocorrosion rate T). In this situation, the total material loss (TML) of the excavator would be computed as the wear (W) occurring during the time operating in dry conditions ($t_{\text{dry duty time}}$), plus the corrosion (C) occurring under wet static conditions ($t_{\text{wet exposure}}$), plus the combined effect of wear and corrosion (T) when operating under wet conditions ($t_{\text{wet duty time}}$).

$$TML = W \cdot t_{\text{dry duty time}} + C \cdot t_{\text{wet exposure}} + T \cdot t_{\text{wet duty time}} \qquad (1.3)$$

This book focuses on tribocorrosion alone, that is the simultaneous action of wear and corrosion, a situation found in different industrial applications. More fundamental aspects of tribocorrosion are discussed in detail in a more comprehensive book [1]. Table 1.1. summarizes tribocorrosion situations in different engineering systems and illustrates how tribocorrosion concepts were applied. The content of the table is based on the literature review carried out by Mischler and Igual Munoz [2].

Table 1.1 Engineering systems in which tribocorrosion was recognized limiting component lifetime

Industrial field	Applications	Examples	References
Biomedical	Artificial joints, dental implants, spine, surgical instruments	Artificial hip joints: modular systems that replace hip joint combining different or the same metal (MoM), ceramics (MoC), polymer (MoP) • Material: CoCrMo, Titanium alloys • Corrosion by: human fluids (saline electrolyte which a complex chemistry containing aggressive ions, water, organic molecules, etc.) • Wear by: cyclic loading Consequence: Detachment of wear particles and release of metal ions in patients with metallic implants or prosthesis	Mischler and Igual Munoz [3]
Marine	Oil and gas pipes, wind turbines, tidal, drilling tools	Anchor systems, chains and connectors: mooring systems allowing for tie up offshore platforms to the sea bottom • Material: steel • Corrosion by: saline environment (aqueous, moisture) and oxygen • Wear by: sliding and fatigue from the ties, wind and waves and boat motions Consequence: Detachment of material and metal dissolution causing lost of mechanical integrity	Wood [4], Harvey [5], Lopez [6]
Civil engineering	Bridges, tunnel boring, mining	Tunnel boring machines: equipment used to excavate tunnels in a huge variety of soil and rock strata by rotating a cutting wheel composed by cutter disks • Material: steel • Corrosion by: soils, salty water and chemical additives of the lubricants • Wear by: impact loading and sliding Consequence: Too often changes of cutters and increase in more than 50% of the total project costs	Espallargas [7]

(continued)

Table 1.1 (continued)

Industrial field	Applications	Examples	References
Transportation	Railway, automotive, maritime transportation, aeronautics	Engine blocks and cylinder heads: engine components for the automotive industry • Material: Al–Si casting alloys • Corrosion by: acids (sulphuric and nitric) formed during low temperature operation of engines (while warming up), when the vapour in combustion gas can condense as water drops and combustion products dissolve in them • Wear by: sliding and fatigue Consequence: Losses of energy and efficiency in the engine performance	Toptan [8]
Manufacturing	Cutting tools	One of the strategies to enlarge the lifetime of cutting tools is based on the use of aqueous cutting fluids because of its lubricant and cooling action • Material: high-speed steel • Corrosion by: aqueous cutting fluids • Wear by: sliding Consequence: Increase in wear rate caused by the acceleration action of corrosive fluids	Kurimoto [9], Fukuda [10]
Energy	Nuclear power plants, wind turbines, oil and gas burners, water turbine	Latch arms: components in pressurized water reactors subjected to mechanical vibrations and high temperatures • Material: stainless steels, cobalt-chromium and zirconium alloys • Corrosion by: pressurized water (formation of thick oxide films) • Wear by: impact, sliding and fretting motion Consequence: Premature failure of nuclear reactor components and high replacing cost	Lemaire [11]

(continued)

Table 1.1 (continued)

Industrial field	Applications	Examples	References
Food	Processing, can manufacturing	Minced meat processing equipment: manufacturing food product equipment consisting of a cabinet with a conveyor screw and perforated disks with knives in sliding contact with the disk • Material: stainless steels • Corrosion by: food (low pH and high chloride concentrations) • Wear by: Consequence: Detachment of wear particles and metal release with the subsequent contamination of food	Jellesen [12]

Tribocorrosion should not be considered only as a thread to engineering systems. It can be also exploited to improve process performances. One example of this is chemo-mechanical polishing (CMP) of integrated circuits where a tribocorrosion situation is on purpose established in order to ensure the quality of the process and to accelerate the production rate [13]. Tribocorrosion process was also used for finishing gears with improved acoustic properties [14].

References

1. D. Landolt, S. Mischler, *Tribocorrosion of Passive Metals and Coatings* (Woodhead Publishing, Lausanne, 2011)
2. S. Mischler, A. Igual Munoz, Tribocorrosion in encyclopedia of interfacial chemistry, 1st edn., in *Edition Surface Science and Electrochemistry*, ed. by K. Wandelt (Elsevier, 2018)
3. S. Mischler, Munoz A. Igual, Wear of CoCrMo alloys used in metal-on-metal hip joints: a tribocorrosion appraisal. Wear **297**, 1081–1094 (2013)
4. R.J.K. Wood, A.S. Bahaj, R. Turnock, L. Wang, M. Evans, Tribological design constraints of marine renewable energy systems. Phil. Trans. R. Soc. A **368**, 4807–4827 (2010)
5. T.J. Harvey, J.A. Wharton, R.J.K. Wood, Development of synergy model for erosion-corrosion of carbon steel in a slurry pot. Tribol. Mater. Surf. Interfaces **1**, 33–47 (2007)
6. A. López, R. Bayón, F. Pagano, A. Igartua, A. Arredondo, J.L, Arana, J.J. González, Tribocorrosion behaviour of mooring high strength low alloy steels in synthetic seawater. Wear **338–339**, 1–10 (2015)
7. N. Espallargas, P.D. Jakobsen, L. Langmaack, F.J. Macias, Influence of corrosion on the abrasion of cutter steels used in TBM tunnelling. Rock Mech. Rock Eng. **48**, 261–275 (2015)
8. F. Toptan, A.C., Alves, I. Kerti, E. Ariza, L.A. Rocha, Corrosion and tribocorrosion behaviour of Al–Si–Cu–Mg alloy and its composites reinforced with B4C particles in 0.05 M NaCl solution. Wear **306**, 27–35 (2013)
9. T. Kurimoto, Improving the performance of aqueous cutting fluids by galvanic-cathodic protection of cutting tools. Wear **127**, 241 (1988)
10. Fukuda et al, Journ. Jap. Wood Research Society **38**, 764–7720 (1992)
11. E. Lemaire, M. Le Calvar, Evidence of tribocorrosion wear in pressurized water reactors. Wear **249**(5–6), 338–344 (2001)

12. M.S. Jellesen, M.O. Hansen, L.O. Hilbert, P. Moller, Corrosion and wear properties of materials used for minced meat production. J. Food Process Eng. **32**, 463–477 (2009)
13. J. Stojadinovic, D. Bouvet, M. Declercq, S. Mischler, Effect of electrode potential on the tribocorrosion of tungsten. Tribol. Int. **4**, 575–583 (2009)
14. K. Oobayashi, K. Irie, F. Honda, Producing gear teeth with high form accuracy and fine surface finish using water-lubricated chemical reactions. Tribol. Intl. **38**, 243–248 (2005)

Chapter 2
Corrosion Basis

Metals are widely used in our society in a variety of applications (engines, cars, airplanes, components in chemical industry, power plants, computers etc.), thanks to their unique combination of mechanical properties and forming easiness. Metals are, however, inclined to react chemically with their environment (corrosion). Rusting of iron is a common example of corrosion leading to material degradation. Corrosion can also have positive effects. For example, aluminium, stainless steels or titanium alloys owe their corrosion resistance to a surface oxide film formed by reaction with water.

Specifically, corrosion can be defined as the "degradation of a material or of its functional properties through a chemical reaction with the environment". Because sometimes corrosion may be desirable (i.e. anodization of aluminium to reinforce the natural oxide film on the surface to increase its corrosion resistance), a broader corrosion definition may be given as corrosion considered "the irreversible interfacial reaction of a material with its environment resulting in the loss of material or in the dissolving of one of the constituents of the environment into the material" [1].

The main corrosion concepts needed for approaching tribocorrosion problems will be introduced in this chapter. The importance of the electrode potential (the difference in potential between a metal and its surrounding environment) as a driving force for corrosion will be highlighted. To do that, the direct relationship between current and reaction rate in a corrosion or tribocorrosion system will be given (which allows quantifying the material loss of a component during operating conditions). Other corrosion relevant tools for understanding tribocorrosion such as the Evans diagrams and passivity will be introduced in the following sections.

© The Author(s), under exclusive license to Springer Nature Switzerland AG 2020
A. Igual Munoz et al., *Tribocorrosion*, SpringerBriefs in Applied
Sciences and Technology, https://doi.org/10.1007/978-3-030-48107-0_2

2.1   Electrochemical Reactions in a Corrosion System

In its simplest form, corrosion of a metal in aqueous solution can be described as the transfer of an atom from the metal (M) into the solution in the form of a dissolved ion (M^{n+}) [1]. Electrochemically this reaction is described by the following reaction designated as oxidation (or anodic reaction) because the metal atom loses n number of electrons.

$$M \rightarrow M^{n+} + n\,e^{-}$$

According to the above reaction, the corrosion of the metal leaves certain number of electrons in the metal that need to be consumed in order to sustain the corrosion reaction. For this, one or more species dissolved in the solution need to undergo reduction at the metal surface (i.e. a chemical reaction involving the consumption of electrons). Usually in aqueous solutions most common reduction (also called cathodic) reactions are the reduction of the proton and/or the reduction of dissolved molecular oxygen as described in the following reactions:

$$O_2 + 2H_2O + 4e^{-} \rightarrow 4OH^{-}$$
$$2H^{+} + 2e^{-} \rightarrow H_2$$

Figure 2.1 shows schematically the reactions taking place on the surface of a corroding metal in aqueous solution.

Most engineering metals used in corrosion application are passive (i.e. stainless steel, Ni-Cr alloys, titanium alloys). They are able to spontaneously form a thin oxide layer (typically few nanometres in thickness) which reduces significantly the corrosion rate. Figure 2.2 schematically shows the oxidation and reduction reactions in a passive material where the overall anodic reaction is the transfer of an atom from the metal (M) into the solution in the form of a dissolved ion (M^{n+}) as already described. However, under passive conditions, this reaction takes place through an intermediate state in which the metal combines with the oxygen from water forming a metal oxide. Further, the metal oxide dissolves and the resulting product, dissolved metal ions (M^{+n}) diffuse into the aqueous environment. The overall reaction rate is limited by the slow transport of ions through the passive film and its interface with the solution.

2.2   Corrosion Rate and Faraday's Law

The amount of reacted metal (i.e. oxidized metal whether in the form of dissolved ions or ion in the oxide lattice) is proportional to the electric charge passing across the electrode/electrolyte interface according to Faraday's law. Therefore, the rate of an electrode reaction (i.e. corrosion rate) is proportional to the current (flow of

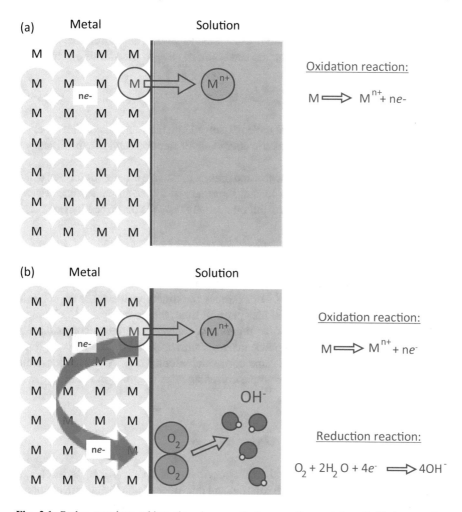

Fig. 2.1 Redox reactions taking place in an actively corroding metal. **a** Oxidation reaction, **b** Oxidation and reduction reaction

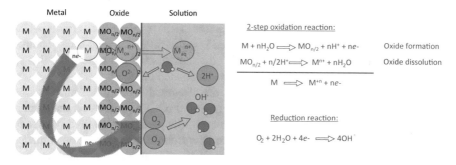

Fig. 2.2 Redox reactions taking place in a passive corroding metal

electrons) flowing through that electrode/electrolyte interface. Equation 2.1 clearly shows that the corrosion of a metal is associated to an internal current between the oxidation site (anode) to the reduction site (cathode) due to transfer of electrons.

$$v'_{corr} = \frac{I}{nF} \tag{2.1}$$

where v'_{corr} is the number of metal moles dissolved into the solution per unit time, I is the current in A, n is the stoichiometric coefficient of electrons and F is the Faraday's number (96485 C/mol).

Corrosion is a surface phenomenon and thus its incidence is better described using the current density (i), that is the current I normalized by the anode surface area A_a (site of oxidation). Thus, the corrosion rate density (v_{corr}) is described by Eq. 2.2.

$$v_{corr} = \frac{i}{nF} \tag{2.2}$$

In engineering, corrosion rates are usually expressed in more practical units than mol per unit time and unit surface. For example, for structures it is more appropriate to consider the linear dimensional loss per unit time, that is mm/year. In chemical engineering the contamination of the liquid is of relevance and thus corrosion rate is usually expressed as released metal mass per unit area and time, that is mg/dm^2 day. All these terms express the same corrosion reaction rate. A conversion table summarizing the factors for determining corrosion rates is shown in Annex A.

2.3 Electrode Potential

Another implication of reaction 1 is that corrosion is characterized by a charge transfer between the metal and the solution and thus depends on the electrical potential difference established at the interface, the so-called electrode potential. The electrode potential can be experimentally determined with respect to a reference electrode or imposed by using electronic devices (i.e. potentiostat/galvanostat). The electrode potential obtained by measuring the voltage between the metal under investigation (working electrode, WE) and a reference electrode (RE) is called open circuit potential (OCP). Alternatively, it can be measured in polarization curves as the potential (called corrosion potential, E_{corr}) at which the current changes sign. Note that in case of equilibrium conditions it can also be calculated based on thermodynamical concepts. The different electrode potentials are discussed in detail in Annex B.

In a corrosion system, the open circuit potential (OCP) depends on the combination of several factors such as solution composition, temperature, hydrodynamic conditions, surface composition and microstructure, strain, contact with other metals and solution inhomogeneity. As an example, OCPs of Ti and its alloys are shown in Table 2.1.

Table 2.1 OCPs of titanium and its alloys as a function of different environmental variables

Influence of	Material	Environment	OCP (V_{SHE})
Acidity	Ti CP	H_2SO_4 0.05 M H_2SO_4 0.5 M	−0.35 −0.69
Chemistry	Ti6Al4V	PBS* PBS + albumin PBS + collagen	−0.18 −0.32 −0.41
Oxidizing agents	Ti6Al4V	Calf serum Calf serum + 0.1% H_2O_2	+0.19 +0.3
Alloying elements	Ti Ti 0.05% Pd	NaCl 4.3 M	−0.51 −0.03

*PBS = Phosphate buffer solution

The OCP can be easily measured using a commercial reference electrode (Fig. 2.3). Note that it exists in different types of reference electrodes, so when giving an electrode potential, the type of reference electrode must be specified (Annex C).

In engineering materials, the so-called galvanic series are typically used to list the OCPs of different metals and alloys in given environments. Note that sometimes the word corrosion potential is used in these series although they were not measured by polarization curves but by simple measurements using a voltmeter (as OCP). As an example, Fig. 2.4 shows OCPs of metal and alloys exposed to seawater.

Table 2.2 shows another example of a galvanic series listing the measured potentials of different materials in soil. Note that depending on the condition of the material (i.e. new or rusted) their relative position (OCP) changes in the series.

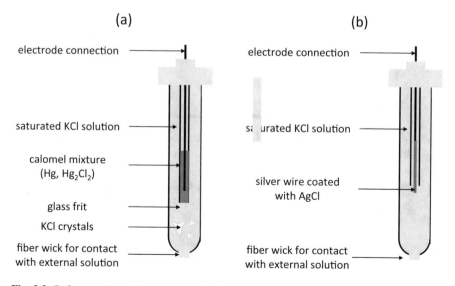

Fig. 2.3 Reference electrodes, **a** saturated calomel electrode, SCE and **b** silver–silver chloride electrode (Ag/AgCl)

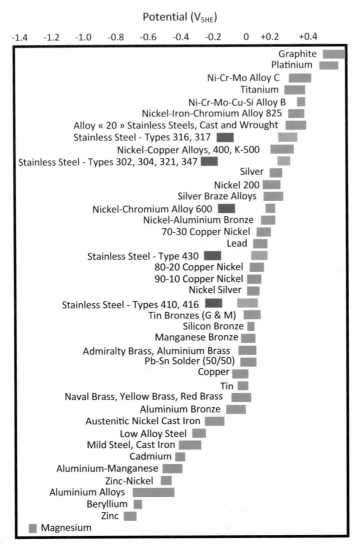

Fig. 2.4 Galvanic series in seawater. In this, green squares correspond to the passive state and red to the active state of the materials

2.4 Evans Diagram: Corrosion Rate Versus Electrode Potential

As already shown above in the listed OCPs for titanium, a metal can present different values depending on the environment. It is therefore necessary to understand how the potential establishes. The Evans diagram is a best way for describing this. They are schematic representations of the anodic and cathodic partial currents as a function of

Table 2.2 OCPs of different materials and conditions in soil

Metals	E (V_{SHE})
Zinc	−0.7 to −0.4
Carbon steel, new	−0.5 to −0.2
Lead	−0.4 to −0.2
Carbon steel, rusted	−0.3 to −0.1
Cast iron, rusted	−0.1 to 0.1
Copper	0.1 to 0.3

potential in a semilogarithmic scale. One typical example of such diagrams is shown in Fig. 2.5. In this figure the evolution of the anodic current density with the potential is shown.

At the reversible potential, no corrosion occurs because the oxidation reaction is in kinetic equilibrium with the reduction of metal ions from the solution according to the following reaction:

$$M^{n+} + n\, e^- \leftrightarrow M$$

The reversible potential is calculated from thermodynamic data. Typically, when increasing the potential above the reversible potential the current (corrosion rate) increases exponentially according to an Arrhenius-type kinetics. Above a critical potential (i.e. passivation potential) a thin metal oxide surface film forms and inhibits corrosion depending on the type of metal and nature of the solution. This corrosion inhibiting effect can be lost at higher potentials (i.e. above the transpassive potential)

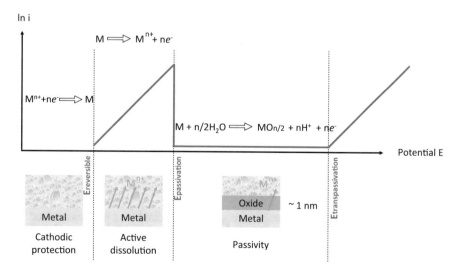

Fig. 2.5 Evans diagram for the anodic reactions of a passive material

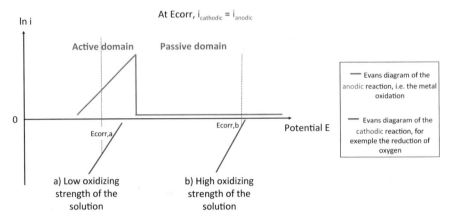

Fig. 2.6 Evans diagram showing the influence of the oxidizing properties of the environment on the corrosion potential and corrosion rate of a passive metal

where local disruption of the film (pitting) or dissolution of the film takes place. Sensitive to the latter effect are specially Cr and Cr-containing alloys where Cr_2O_3 forms at low potentials providing very effective passivity but becomes soluble when oxidizing to CrO_3 at potentials above the transpassive potential.

The electrode potential spontaneously attained by a metal in a given solution depends on the oxidizing strength of the solution (nature and concentration of the oxidizing agent, catalytic properties of the surface) and it corresponds in principle to the experimentally determined corrosion potential (E_{corr}) and open circuit potential (OCP). At E_{corr} the reaction rates of the metal oxidation (positive current) and of the reduction reaction (negative current) occur at the same rate. Thus, when plotting in an Evans diagram both the anodic and the cathodic curves, the E_{corr} will correspond to the potential at which both curves equilibrate (see Fig. 2.6). Thus, depending on the oxidizing strength of the solution, the potential can lay in the active domain or in the passive domain, consequently leading to different corrosion rates as shown in the Fig. 2.6. Note that the potential is not the only determining parameter for a metal to be in an active or passive state, but other actions such as mechanical abrasion may cause the material to switch from passive to active (see Chap. 4).

2.5 Corrosion Types

In practice eight types of corrosion are commonly classified as summarized in Table 2.3 [2].

Table 2.3 Types of corrosion

Type	Definition	Scheme	Example/image
Uniform	Material loss that occurs with the same rate all over the entire surface		Brass valve in a heat exchanger component
Galvanic	When two different metals in electrical contact (physical or through the electrolyte) are exposed to a corrosive environment leading to corrosion of the less noble material		Water pipe junction in a saline atmosphere
Crevice	Localized attack occurring in confined spaces where the oxidant is not available		Stainless steel flange exposed to a chloride containing solution
Pitting	Localized attack in passive materials due to imperfections and presence of aggressive anions (i.e. Cl^-)		Localized pits on a heat treated stainless steel

(continued)

Table 2.3 (continued)

Type	Definition	Scheme	Example/image
Intergranular	Selective attack in the grain boundaries of the material		Austenitic Cr-Ni-Mo stainless steel after operating in an urea production line
Selective leaching	Selective attack of the less noble alloying element occurring in multiphase alloys (i.e. brass)		Bronze (Cu-Sn) material operating in offfshore
Erosion corrosion	Material loss due to the mechanical action of hard particles and/or droplets in corrosive electrolyte		Outer and inner parts of a copper pipe of a heat exchanger (arrows indicate flow direction)
Stress corrosion cracking	Enhanced material loss due to acting mechanical forces in corrosive environment		Stainless steel under subsea conditions

Acknowledgement The authors would like to thank Cristian Torres (NTNU), Roy Johnsen (NTNU) and Claes Olsson (Upssala University) for providing some of the images of the corrosion types shown in Table 2.3.

References

1. D. Landolt, *Corrosion and Surface Chemistry of Metals* (EPFL Press, Lausanne, 2007)
2. M.G. Fontana, N.D. Green, Corrosion Engineering (McGraw Hill Series, 1978)

Chapter 3
Tribology Basis

Tribology deals with the study of the technological-relevant phenomena of wear, friction and lubrication. Tribological situations arise when two components are loaded and move against each other. Friction is the force opposing or blocking the relative motion between two bodies in contact. In some applications friction is desired, for example in the grasping of objects, or for braking. In other cases, as in tobogganing, skiing or displacement of objects friction needs to be minimized. Lubrication consists of separating two contacting bodies by a deformable (low shear strength) film that reduces friction and/or wear. Well-known examples of lubricants are oils and greases. Wear is the progressive material loss from contacting bodies in a tribological system.

These tribological phenomena result from complex surface interactions of bodies in relative motion and involve the interplay of mechanical, material, physical and chemical factors. This chapter introduces the basic concepts of contacting surfaces, friction, wear and lubrication relevant for tribocorrosion.

3.1 Surfaces and Contact Mechanics

When loading two bodies against each other, the contacting surfaces can undergo, depending on prevailing stresses, elastic (reversible) deformation and, in addition, when the load exceeds a critical threshold, plastic (permanent) deformation. Plastic deformation is one of the main factors causing wear and friction of metals. In this section the contact mechanics concepts that allow to assess contact stresses and the occurrence of plastic deformation are introduced.

3.1.1 Elastic Contact

The elastic deformation and the resulting contact pressure of perfectly smooth surfaces in a non-conformal contact can be determined by Hertz theory [1, 2]. Hertz assumed a frictionless contact, smooth surfaces, elastic deformations and small deformations. For roller bearings, gears and cam-followers this may be the case when the surfaces have been run in. As an example, the Hertz equations to calculate the contact pressure and the area of contact for a ball-on-flat geometry are given in Fig. 3.1. For other types of geometries, the reader is referred to [3].

$$a = \left(1.5\, F_n R' / E'\right)^{1/3} \tag{3.1}$$

$$P_{\max} = 3\, F_n / 2\pi\, a^2 \tag{3.2}$$

$$P_{\text{average}} = F_n / \pi\, a^2 \tag{3.3}$$

where a is the radius of the contact area [m], F_n is the normal load [N]; P is the contact pressure (Hertzian stress) [Pa]; E' is the reduced Young's modulus [Pa]; and R' is the reduced radius of curvature [m].

The reduced Young's modulus is defined as:

$$\frac{1}{E'} = \frac{1}{2}\left[\frac{1 - \nu_A^2}{E_A} + \frac{1 - \nu_B^2}{E_B}\right] \tag{3.4}$$

where ν_A and ν_B are the Poisson's ratios of the contacting bodies A and B, respectively, and E_A and E_B are the Young's moduli of the contacting bodies A and B, respectively.

The reduced radius of curvature for a ball-on-flat contact is defined according to this theory as

$$\frac{1}{R'} = \frac{1}{R_x} + \frac{1}{R_y} = \frac{1}{R_{Ax}} + \frac{1}{R_{Ay}} = \frac{2}{R_A} \tag{3.5}$$

Fig. 3.1 Elastic deformation of ball-on-flat contact under a load F_n

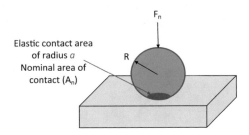

Table 3.1 Contact mechanics parameters for a ball-on-flat contact with different kind of ball materials of 5 mm in diameter sliding against a rigid one at 10 N of applied normal load

Ball Material	Elastomer	Polymer	Metal	Ceramic	Unit
E Modulus	0.02	1	200	500	GPa
Poisson ratio	0.5	0.5	0.3	0.3	
Radius of contact area	1.121	0.304	0.057	0.043	mm
Average pressure	3	34	995	1740	MPa
Yield strength (σ_Y)	10	20	350	350	MPa
Av. pressure/σ_Y	0.3	1.7	2.8	5.0	

Table 3.1 shows typical values of contact mechanics parameters for a ball-on-disk contact (Fig. 3.1) with a 5 mm diameter ball made of different material types (elastomer, polymer, metal or ceramic) at an applied normal load of 10 N against a rigid plane calculated with the Hertz formalism. In the table, the yield strength of the different materials is also included. Considering that the onset of plastic deformation occurs approximately when the average contact pressure exceeds the yield strength (σ_Y) of the material (Eq. 3.6), it is possible to see how the nature of materials influences the onset of plastic deformation. In the examples, except the elastomers, all materials will yield.

$$P_{\text{average}} > 1.1\,\sigma_Y \tag{3.6}$$

3.1.2 Plastic Contact

Up to now, we have been dealing with ideal smooth surfaces (nominal area of contact A_n, Fig. 3.1) but real surfaces are rough and therefore only asperities will get in contact. Those asperities will plastically yield as a consequence of high local pressure (Fig. 3.2).

Real surfaces are characterized by its topography. As an example, Fig. 3.3 shows a real surface with its surface profile. Details of how to measure and characterize surface topography are given in Annex D.

The real area of contact (A_r) can be calculated by the sum of the area of the junctions (a_i) as shown in Fig. 3.4 and defined by the following equations:

$$a_i = f_i/H \tag{3.7}$$

$$A_r = F_n/H \tag{3.8}$$

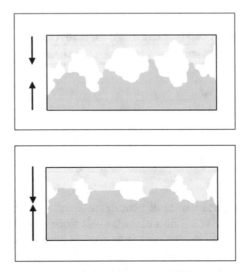

Fig. 3.2 Scheme of the contact of real surface at the asperity junctions

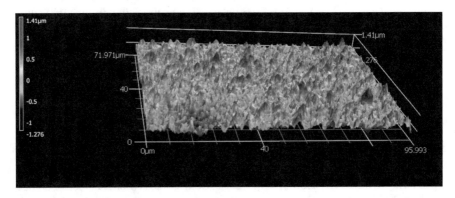

Fig. 3.3 3D-image of a copper surface

Fig. 3.4 Real contact area (A_r) which corresponds to the sum of the individual contact asperities (a_i)

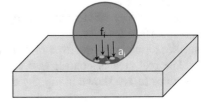

where f_i is the normal load carried out by a single area of a junction and H is the surface hardness. In tribology surface hardness is an important parameter that determines the real contact area. Measurements of hardness of the bulk and worn areas of the material after tribological tests can be a useful parameter for understanding the causes of damage but it is also a needed parameter in using available models for wear predictions such as wear of CoCrMo alloys [4].

3.2 Friction

Friction is the force resisting relative motion between two solids in contact when an external tangential force is applied (Fig. 3.5).

The friction force is not an intrinsic material property of the contacting materials, but rather a response of the contact system as a whole (e.g. contacting materials, presence of oxides or lubricants and atmospheric conditions). Czichos describes a tribo-system by four elements [5]:

- The two contacting bodies, including the geometry, surface topography and mechanical properties.
- The environment
- The lubricant

Moreover, friction depends on several other parameters:

- Contact forces (normal and tangential)
- Type of relative motion (static, rolling, sliding, etc.)
- The velocity of the relative motion.

The investigation of the mechanisms of friction was first done by Leonardo da Vinci (1452–1519), and later worked by Amonton (1663–1705) and Coulomb (1736–1806). These works resulted in three general laws of friction [2].

1. Friction force is directly proportional to the applied load.

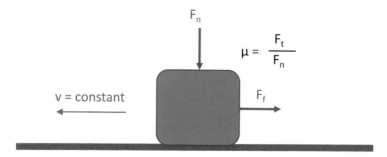

Fig. 3.5 Sketch of friction in which a frictional force (F_f) is needed to keep an object sliding at a constant velocity

2. Friction force is independent of the apparent area of contact.
3. Friction force is independent of the sliding speed.

According to the first law, one can define the coefficient of friction as the ratio between the friction force and the applied normal force as shown in Eq. (3.9).

$$\mu = \frac{Ft}{Fn} \tag{3.9}$$

where F_t is the friction force (N) and F_n is the applied normal force (N).

These three laws of friction were deducted by empirical observations of sliding pairs constituted by metals and ceramics and not derived by fundamental aspects of material interactions. As a consequence, although still describing friction in many cases with sufficient accuracy these laws failed to explain the frictional behaviour of certain polymers. For these reasons, the three laws should be rather considered as rules.

The implications are that friction is a system-dependent parameter rather than a material property, thus varying as a function of parameters such as the environment or the contact configuration. Table 3.2 confirms this characteristic by showing typical values of the coefficient of friction (COF). Note that the given COF values are only tentative ones, obtained under certain experimental conditions.

Several models have been proposed for describing friction mechanisms. Most of the current sliding friction theories are based on the work carried out by Bowden Tabor [6] who assumed that frictional force arises from an adhesion force produced

Table 3.2 Typical COF values of different tribological systems

			COF
Influence of sliding partner (X)	X / Steel 1032	X = Al6061 T6	0.38
		X = Copper	0.25
		X = Steel 1032	0.23
		X = Teflon	0.07
Influence of contact configuration	Al 6061T6 / Ti6Al4V		0.38
	Ti6Al4V / Al 6061 T6		0.29
Influence of environment	Fe / Fe	Vacuum	>4 (seizure)
		10^{-3} mbar O_2	1.5
		1 mbar O_2	0.4
		Oil film	<0.1

at the asperity junctions (real areas of contact) and a deformation force needed to plough the asperities of the harder surface through the softer surface [2]. Nowadays it is known that these two contributions cannot be considered separately and that the reason for a metal flowing plastically or not is determined by a yield criterion (i.e. Tresca's or von Misses yield criterion).

3.3 Friction, Contact Mechanics and Plastic Deformation

When a metallic contact is in motion, wear is mainly due to deformation of the contacting asperities. At small normal loads, the response is perfectly elastic and reversible if none of the contacting metals yields. If the asperities yield, the structure can elastically shakedown in case the work hardening increases the yield strength above the applied stress. The maximum load for which it happens is named the elastic shakedown limit. Above this limit, the material undergoes reversible cyclic plastic deformation. In this case failure is expected by a low-cycle fatigue mechanism. Finally, if the load is increased above the plastic shakedown limit, ratcheting occurs, meaning accumulation of plastic deformation at the surface at each cycle. Failure of the material occurs when the accumulated strain exceeds a critical value.

The above-described phenomena can be represented in a shakedown map [7], which shows the influence of the coefficient of friction and the load factor ($p_{max}/0.5\sigma_Y$) on the shakedown limits (Fig. 3.6). There is a critical value of the coefficient of friction at about 0.25 in the shakedown map. Below this value, subsurface stresses control the elastic shakedown, and above this value plastic deformation of the surface controls the shakedown limit and decreases with $1/\mu$.

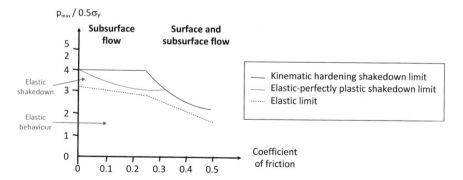

Fig. 3.6 Shakedown map

3.4 Wear

Wear is defined as the progressive loss of material at the surface of a contacting body occurring as a result of relative motion with a counter-body. It can also be defined as the removal of material from a solid surface due to the effect of mechanical action. Wear occurs in a wide variety of contacts and vary from one single mechanism to a combination of several wear mechanisms. Figure 3.7 shows an example of a worn surface of an AISI 316L after tribocorrosion test in H_2SO_4. In the figure, the damaged surface shows the simultaneous consequences of plastic deformation, subsurface cracking and wear debris generation. It demonstrates how different wear phenomena may take place simultaneously.

As with friction, wear resistance is not an intrinsic property of the materials, but rather the response to the tribo-system as a whole. However, as wear involves the loss or displacement of material, properties such as hardness, fracture toughness and chemical stability are more closely linked to the wear resistance of certain tribological system than they are for frictional response.

3.4.1 Wear Mechanisms

Different wear mechanisms can be distinguished depending on the relationship between the loading conditions, contact pressure and material properties:

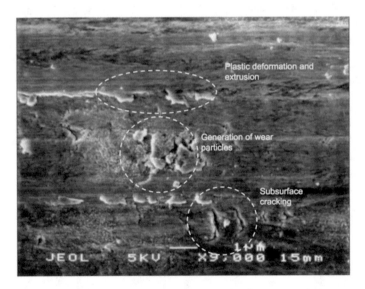

Fig. 3.7 SEM image of a stainless steel worn surface with the indication of different degradation mechanisms

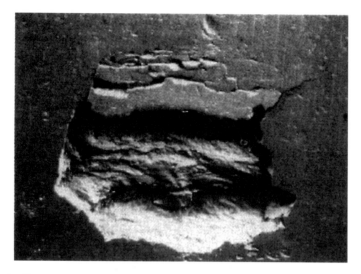

Fig. 3.8 Spalling off of metal particles after large number of loading cycles. Form [5]

- If the average contact pressure is lower than the yield strength of the contacting material only elastic (reversible) deformation occurs. Fracture can nevertheless occur through the mechanism of fatigue when the material is cyclically loaded as in the case of impacts or rolling. Cracks can develop and propagate under repeated loading even if the generated stress remains below the yield strength of the material. Once a crack reaches a critical length or the surface then a fragment of material can detach as shown in Fig. 3.8. This kind of mechanism is called **fatigue wear**.

  Even sliding surface under constant normal force experiences alternative loading. Suppose for this that a perfectly rigid sphere elastically indents and slides against a flat elastic surface. Due to the motion, the surface contacting the leading edge of the ball experiences compression while the surface at the trailing edge experiences traction. This change in stress state can in case of repeated passes initiate fatigue cracks that grow at each loading cycle. Note than at a microscopic scale this kind of process can occur at each asperity contact in case of non-perfectly smooth surfaces.

- When the average contact pressure exceeds the yield strength of the material then plastic (irreversible) deformation occurs and the material yields. This plastic deformation in the contact area has several consequences. First, the generation of dislocations and other microstructural defects in the deforming material work harden the material and as a consequence its yield strength increases. This hardening may in certain case be sufficient to set back the elastic stress below the yield strength and block further plastic deformation. If not, then wear can occur by cutting or ploughing.

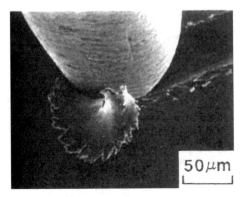

Fig. 3.9 Metal cutting directly forms wear particles (abrasion). From [8]

Sharp-edged sliding bodies can directly cut sections of the softer sliding material due to the high strain generated at indenter edges. This wear mechanism is called cutting or, referring to surface asperities acting as indenters, **micro-cutting** (Fig. 3.9).

In case of blunt asperities or sliders, plastic deformation does not directly result in the chipping off material as in the case of cutting. Rather, the material in the contact zone is pushed apart by the moving indenter (ploughing). As a consequence, strain continues to accumulate with increasing number of sliding passes into the deforming material. Eventually, the accumulated strain locally exceeds the critical threshold that the material can withstand (ratchetting threshold). In this case the material breaks down and generates wear particles through the so-called **(micro) ploughing** mechanisms (Fig. 3.10).

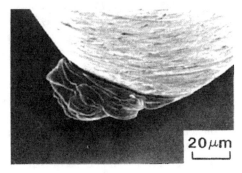

Fig. 3.10 Microploughing occurs in two phases, first strain accumulation during repeated passes (no wear) tales place and secondly, by breaking (generation of wear particles) when the accumulated strain is higher than a critical strain. From [8]. Note that in case of micro-cutting ejection of wear particles is immediate while in case of microploughing an incubation time is required to accumulate the strain necessary to release a wear particle

3.4.2 Wear Quantification

Prediction of wear through theoretical formalisms is available in the literature. Indeed, Meng and Ludema [9] published a paper where they found more than 182 equations for wear published between 1955 and 1995. However, the most well-known and widely used is the Archard approach [10] which establishes a proportionality between wear, the real area of contact and the sliding distance in case of plastic contact (Eq. 3.10)

$$V_{\text{tot}} = K \, L \, F_n / H \qquad (3.10)$$

where V_{tot} is the wear volume loss (mm^3), K the non-dimensional wear coefficient, L is the sliding distance (m), F_n the normal load (N) and H the hardness (MPa) of the softer surface.

Wear is proportional to the applied normal force (F_n), the sliding distance (L) and the real area of contact (A_r). Therefore, several ways of defining wear rate are:

$$v_{\text{wear}} = V_{\text{tot}}/L \left(\text{volume loss per unit of sliding distance} \left[\text{mm}^3/\text{m}\right]\right) \qquad (3.11)$$

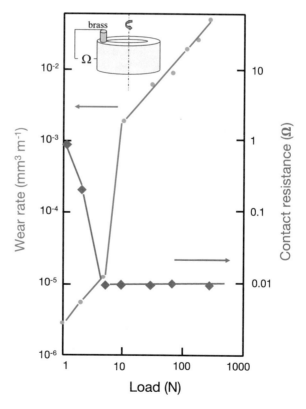

Fig. 3.11 Wear transition for a leaded brass pin sliding against a hard Stellite ring. From [11]

$$v_{\text{wear}} = V_{\text{tot}}/(L\ F_n)\left(\text{wear coefficient}\left[\text{mm}^3/\text{m N}\right]\right) \qquad (3.12)$$

$$v_{\text{wear}} = V_{\text{tot}}H/(L\ F_n)(\text{dimensionless wear coefficient}) \qquad (3.13)$$

These expressions do not take into account chemical (oxidation, corrosion, etc.), metallurgical (hardening, etc.) or physical (temperature, particles, etc.) transformations that may occur during a tribological test. This is the case, for example, of wear of brass sliding against a Stellite in air with variable load (Fig. 3.11). In this system, at low loads, wear rate increases with the normal load according to Eq. 3.10. At higher loads, again, wear follows Archard equation, while at intermediate loads (5–10 N) there is a sharp increase in wear rate by a factor of about 100. This wear transition is associated to the presence of a thin oxide film at the surface of the brass. In case of low loads, the local contact pressure at asperities contacts is insufficient to penetrate the oxide film and this is the reason why higher contact resistance is measured.

Experimentally, material loss can be measured with different techniques, for example, ex situ and in situ. The most typical one is profilometry. With this technique, one can determine the material removed or displaced in a contact by measuring the height profile across the wear track. Figure 3.12 shows two schematic illustrations of how to quantify displaced or removed material from a wear track. The total material loss can then be quantified by determining the worn area from the profile and multiplying it by the length of the wear track.

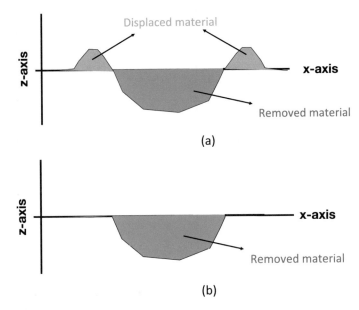

Fig. 3.12 Schematic illustration of wear tracks with **a** the displaced and removed material and **b** only removed material

Other techniques such as mass loss (gravimetric) can also be used if the amount of material lost is large enough. The disadvantage of gravimetric techniques is that a mass increase might also appear if samples get contaminated or if corrosion products are deposited on the surface. In situ techniques for measuring wear have appeared in recent years such as the real-time impingement and radio nuclide [12].

3.5 Lubrication

Lubrication is a term used when a low shear strength layer of gas, liquid or solid is formed/placed between two sliding surfaces. This layer typically separates the surfaces, facilitates low friction sliding and/or reduces wear. Depending on the velocity and conditions of the lubricated contacts, hydrodynamic, elastohydrodynamic (mixed) or boundary lubrication regimes are found. A way of graphically showing the three lubricated regimes found in lubricated contacts is the use of Stribeck curves [13, 14]. This curve typically represents either the coefficient of friction, the wear or the film thickness versus a relationship of physical parameters such as the sliding velocity (v_s), the viscosity of the lubricant (η) and the normal load (Fn). An example of a typical Stribeck curve obtained in an oil lubricated system where the different lubricating regimes achieved by changing the speed of the test is shown in Fig. 3.13.

The lambda ratio (λ) is a parameter that can be used for determining the lubricating regime at which a system is operating. This ratio can be found in Eq. 3.14:

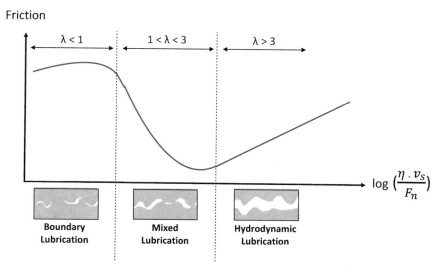

Fig. 3.13 Variation of friction as a function of lubricated surfaces (Stribeck curve)

Table 3.3 Lubricating regimes

λ value	Type of lubricating regime	
<1	Boundary	Wear is expected and true contact between surfaces occurs
1–3	Mixed	Some plastic deformation of surface asperities take place leading to some minor wear
>3	Hydrodynamic	Full separation of surfaces occurs with no wear

$$\lambda = \frac{h_0}{\sqrt{\left(R_{qA}^2 + R_{qB}^2\right)}} \tag{3.14}$$

where h_0 is the minimum film thickness, R_{qA} and R_{qB} are the root mean square roughness of the surfaces in contact. R_q corresponds to the root mean square roughness and is the geometric average value of the profile departure from the mean line within a sampling length.

The minimum fluid film thickness in elastohydrodynamic conditions can be determined by the Hamrock–Dowson equation [15]:

$$h_{\min} = 2.8 \left(\frac{u\eta}{E'R'}\right)^{0.65} \left(\frac{F_n}{E'R'^2}\right)^{-0.21} R' \tag{3.15}$$

Table 3.3 shows the different lubricating regimes depending on the range of λ value.

Tribocorrosion systems are typically lubricated by water-based solutions of low viscosity which implies typically boundary lubricating conditions. However, in the case of very smooth surfaces such as hip joints, hydrodynamic lubrication can be achieved [4].

References

1. H. Hertz, Über die Berührung fester elastischer Körper (On the contact of elastic solids). J. reine und angewandte Mathematik **92**, 156–171 (1882)
2. I.M. Hutchings, *Tribology: Friction and Wear of Engineering Materials* (CRC Press, 1992)
3. G.W. Stachowiak, A.W. Batchelor, *Engineering Tribology* (Elsevier Butterwoth-Heinemann, Burlington, 2005)
4. S. Cao, S. Guadalupe Maldonado, S. Mischler, Tribocorrosion of passive metals in the mixed lubrication regime: theoretical model and application to metal-on-metal artificial hip joints. Wear **324–325**, 55–63 (2015)
5. H. Czichos, *Tribology. A System Approach to the Science and Technology of Friction, Lubrication and Wear* (Elsevier, Amsterdam, 1978)
6. F.P. Bowden, D. Tabor, *The Friction and Lubrication of Solids Part I.* (The Clarendon Press, Oxford, 1950), pp. 321–327
7. K.L. Johnson, Contact mechanics and the wear of metals. Wear **190**(2), 162–170 (1995)
8. K. Kato, Wear model transitions. Scripta Metall. **24**, 815–820 (1990)

9. H.C. Meng, K.C. Ludema, Wear models and predictive equations: their form and content. Wear **181–183**, 443–457 (1995)
10. J.F. Archard, Contact and rubbing of flat surfaces. J. Appl. Phys. **24**, 981–988 (1953)
11. W. Hirst, J.K. Lancaster, J. Appl. Phys. **27**, 1057–1065 (1956)
12. M. Dienwiebel, M.I. De Barros Bouchet, *Advanced Analytical Methods in Tribology* (Springer, 2018)
13. R. Stribeck, Die wesentlichen Eigenschaften der Gleit- und Rollenlager (The basic properties of sliding and rolling bearings), Zeitschrift des Vereins Deutscher Ingenieure, 1903, Nr. 36, Band 46, p. 1341–1348, p. 1432–1438 and 1463–470
14. H. Czichos, K.-H. Habig, *Tribologie-Handbuch (Tribology Handbook)*, 2nd edn. (Vieweg Verlag, Wiesbaden, 2003)
15. B. Hamrock, D. Dowson, Elastohydrodynamic lubrication of elliptical contacts for material of low elastic modulus I-fully flooded conjunction. J. Lubr. Technol. **100**(2), 236–245 (1978)

Chapter 4
Tribocorrosion Phenomena and Concepts

Material removal from surfaces sliding in a corrosive environment occurs by two fundamental mechanisms: mechanical detachment of base material particles (i.e. debris) and emission or ejection of chemically reacted base material particles. These mechanisms take place at different surface zones of the tribo-system in passive materials, as illustrated in Fig. 4.1.

Corrosion occurs in the mechanically unaffected surface of the material (which is typically covered by an oxide film in case of passive metals (I in Fig. 4.1)). Although not directly mechanically loaded, this zone is subjected to changes in electrode potential and corrosion mechanisms induced by the galvanic coupling with the worn surface (III in Fig. 4.1), shifting the OCP towards lower values due to cyclic passive film removal (see Chap. 6, Sect. 6.1). V_{corr} quantifies the amount of lost material as a consequence of the corrosion phenomenon occurring in zone I.

In the contact area (II in Fig. 4.1), wear results in the detachment of metal particles (*mechanical wear, V_{mech}*) as well as possible chemical activation of the metal surface. Chemical activation may result from surface roughening, removal of adsorbed species on the surface or removal of passive films. The most common materials used in tribo-corrosion systems are passive metals that are mechanically activated by the removal of the passive film. The depassivated area (III in Fig. 4.1) exposed to the electrolyte solution corrodes at a much higher corrosion rate (*wear accelerated corrosion, V_{chem}*) until the passive film forms again (repassivation). This depassivation/repassivation process (wear accelerated corrosion) is kept by the repeated action of the counterpart on the wear track.

Due to the concurrence of these phenomena, the total material loss (V_{tot}) in a tribocorrosion situation can be expressed as the sum of the following contributions:

$$V_{tot} = V_{mech} + V_{chem} + V_{corr} \qquad (4.1)$$

In case of passive materials, the V_{corr} contribution is much lower than the other two contributions and it is therefore typically neglected.

© The Author(s), under exclusive license to Springer Nature Switzerland AG 2020   35
A. Igual Munoz et al., *Tribocorrosion*, SpringerBriefs in Applied
Sciences and Technology, https://doi.org/10.1007/978-3-030-48107-0_4

Fig. 4.1 Scheme of a
tribocorrosion contact and
the three different surface
zones distinguished in a
tribocorrosion situation. The
system is immersed in an
electrolyte solution. From [1]

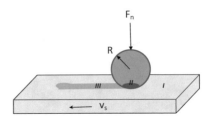

4.1 Effect of Corrosion on Wear

Corrosion can influence wear in different ways. Mechanical weakening of the material can occur due to localized corrosion or selective dissolution of one or more of the metal components. For example, selective corrosion of grain boundaries weakens the cohesion of the surface grains and thus potentially promotes wear (Fig. 4.2). Pits formed by chloride attack in seawater can act as crack nucleation sites promoting fatigue wear of the material [2].

Metal alloys can undergo selective dissolution of one or more of the alloying elements. A well-known example is the dezincification [3] of brass (Cu–Zn alloy) in natural water, resulting from the preferential dissolution of the less noble component (Zn). As a consequence, the surface becomes porous and highly enriched in Cu and therefore a change in the mechanical properties is expected.

In some cases, surface corrosion products can act as boundary lubricants and protect mechanically the underlying material. For example, Drees et al. found that the favourable wear behaviour of TiN coating in phosphoric acid solutions was due to the precipitation of phosphate surface films [4]. In addition, oxide films may influence the surface mechanical response of metals. This effect is known as Roscoe effect [5]. Zn was found to behave perfectly plastic when scratched in the absence of naturally grown oxide film, but under identical experimental conditions the deformation of

Fig. 4.2 Weakening of a
surface by selective
corrosion of grain
boundaries. Courtesy of
wikicommons

720X ———— 20 μm

oxidized Zn was restricted and twins appeared on the scratched surface. This effect is thought to play a major role in the tribocorrosion of passive metals, in particular on mechanical wear.

4.2 Effect of Wear on Corrosion

Passive metals are particularly sensitive to faster corrosion when wear occurs simultaneously. This is due to the cyclic removal of the passive film by abrasion followed by its regrowth (depassivation/repassivation mechanism). The regrowth of the passive film requires a certain time in which a significant dissolution of the metal takes place. Depassivation is the increase of corrosion kinetics due to the removal of the passive film, while repassivation is considered as the capability of a passive material to reform the damaged passive layer. This mechanism can increase the corrosion rate by several orders of magnitude [6]. Figure 4.3 schematically illustrates this phenomenon.

Wear-accelerated corrosion can be in situ quantified by measuring the current during rubbing, as it will be explained in Sect. 6.1. The passivation charge density, Q_p (mC/cm^2) is the electrochemical parameter that corresponds to the charge density needed to passivate a unit of a bare metal surface (from which the passive film has been removed). A review of how to measure Q_p has been carried out by Mischler et al. [6], which summarizes the most commonly used electrochemical techniques such as current transient analysis under potentiostatic conditions.

Interestingly, the corrosion rate of active materials is comparatively only marginally affected by wear [7] and therefore, non-passive materials appear as more appropriate candidates than passive metals for tribocorrosion applications. However, one needs to consider the overall corrosion rate of the metal which is typically much larger in case of active metals.

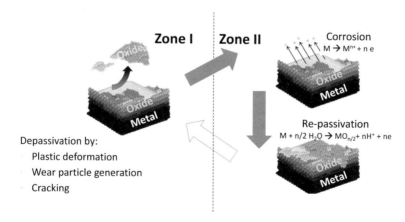

Fig. 4.3 Cyclic depassivation/repassivation mechanism (wear accelerated corrosion) in a tribocorrosion system

4.3 Third Bodies and Transfer Films

During sliding, particle detachment due to wear can generate third bodies composed of wear particles trapped between the two contacting bodies. These particles (third bodies) can remain in the contact, can leave the contact, can be oxidized, can smear on one or the two surfaces (i.e. create a transfer film), and so on [8]. A sketch of this phenomenon is shown in Fig. 4.4.

The build-up of third bodies fundamental modifies the contact conditions that can no longer be described by just considering the two initial bodies. The wear must then be described as a mass flow between the first, second and the third body [9].

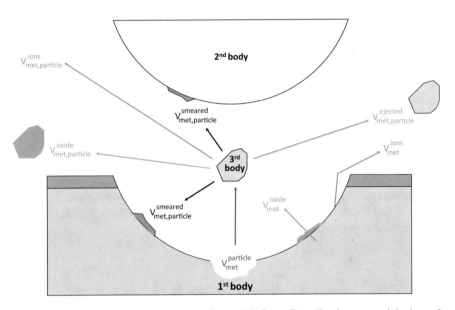

Fig. 4.4 Sketch of third body situation with the particle flows. From [book, anna-nuria's chapter], where $V_{\text{met}}^{\text{particle}}$ is the material detached as solid metal particles due to abrasion, adhesion or delamination, $V_{\text{met}}^{\text{ions}}$ is the material removed as metal ions dissolved in the electrolyte, $V_{\text{met}}^{\text{oxide}}$ is the metal oxidized to form the passive film, $V_{\text{met,particle}}^{\text{ejected}}$ are metal particles ejected from the contact, $V_{\text{met,particle}}^{\text{ions}}$ are metal particles oxidized in the form of ions, $V_{\text{met,particle}}^{\text{oxide}}$ are metal particles oxidized in the form of oxide, and $V_{\text{met,particle}}^{\text{smeared}}$ are metal particles transferred back to the metal surface by smearing

4.4 Models

Modelling allows developing a comprehensive understanding of any phenomena but in the tribocorrosion field, it specially allows to reduce complex and expensive experimental programmes as well as giving directions for the design of tribocorrosion materials, materials surface treatments or equipment design. Different tribocorrosion models have been proposed for different situations that are summarized in Table 4.1 and recently reviewed by Cao et al. [10].

The first tribocorrosion model was proposed by Uhlig in 1954 [11] for a dry fretting corrosion system in which degradation was described in terms of mechanical wear and chemical wear resulting from the removal of the air formed oxide film. Forty years later, this approach was extended by Mischler et al. [6] for a sliding tribocorrosion system where depassivation was considered to occur by plastic deformation of asperities (Eq. 4.2).

$$I_a = R_{dep} Q_p = k_a v_s \left(\frac{F_n}{H} \right)^{0.5} Q_p \qquad (4.2)$$

where I_a is the anodic current (A), R_{dep} is the depassivation rate, Q_p is the passivation charge density (mC/cm^2), k_a is a constant, v_s is the sliding velocity (mm/s), F_n is the

Table 4.1 Existing tribocorrosion models

Model [Refs.]	Predicted value	Materials	Conditions	Environment	Features
Mechanistic Uhlig [11]	V_{chem}	Steel	Fretting	Dry	Atmospheric corrosion
Mechanistic Mischler [6]	V_{chem}	Stainless steel	Sliding	Acid and neutral aqueous solutions	Wear-accelerated corrosion
Fatigue-wear Jiang [13]	V_{tot}	SS	Sliding-wear	Seawater	Combines wear and corrosion
Tribocorrosion-fatigue Von der Ohe [15]	V_{tot}	SS	Sliding wear combined with fatigue	Seawater	Combines fatigue wear and corrosion
Galvanic Vieira [12]	Potential	Al alloys CoCr Ti SS	Sliding	Neutral aqueous solutions	Galvanic coupling between worn and unworn
Third body Guadalupe [9]	V_{tot}	Stellite	Sliding with formation third bodies	High temperature	Wear as material flow through third body
Mechanistic with lubrication Cao [17]	V_{tot}	CoCrMo	Sliding	Various aqueous solutions	Combines lubrication, wear and corrosion

normal force (N) and H is the microsurface hardness (Pa). Practical details of how to determine and use Q_p are given in the example of chemo-mechanical polishing of tungsten case study 2 in Sect. 8.2. This wear model was, for example, successfully applied to rationalize the effect of oxidizing agents in CMP slurries (see case 2, Sect. 8.2).

Equation 4.1 is mainly applicable at imposed passive potentials where anodic current (I_a) can be measured by a potentiostat and can be used to quantify chemical wear (V_{chem}). In tribocorrosion tests carried out at OCP (see Sect. 6.1 in Chapter 6), a galvanic couple is established between the worn (anode) and the unworn (cathode) areas, and the anodic current (I_a) cannot be directly measured; therefore the use of Eq. 4.1 for quantifying chemical wear becomes difficult. However, Viera et al. [12] proposed a model for quantifying the chemical wear of passive materials tested at OCP known as the "galvanic coupling" model (Eq. 4.3).

$$E_c = E_{corr} + a_c - b_c \log i_a \qquad (4.3)$$

where E_c is the potential measured during rubbing, E_{corr} is the corrosion potential of the material, a_c and b_c are the Tafel coefficients and i_a is the anodic current density (I_a divided by the worn area, anode). A practical example on how to use this equation to determine wear in hip joint implants is given in Chapter 8, Sect. 8.4.

By using Eq. 4.3, it is possible to estimate i_a by only measuring the OCP in a tribocorrosion experiment and determining the cathode kinetics by a Tafel extrapolation of the linear part of the cathodic branch in an independent and simple electrochemical test such as a cathodic polarization curve. In case study 4 (Sect. 8.4 in Chapter 8) the wear-accelerated corrosion rate of metal-on-metal hip joints was determined from OCP measurements in an electrochemically instrumented simulator (see Fig. 5.5) and by using this model.

A model for predicting wear in a sliding tribocorrosion system was proposed by Jiang et al. [13, 14]. This model assumed that a fatigue process was the responsible for the material loss. A low cycle fatigue wear formalism was used, making fatigue crack generation and propagation responsible for the increase in corrosion kinetics. This approach was further developed later for a sliding tribocorrosion system with combined external fatigue [15].

During tribocorrosion, build-up of solid third bodies separating the two initial bodies in contact is often observed. These third bodies may have different chemical composition than the initial contacting bodies, and therefore wear is not simply a first body reaction to a mechanical action, but involves a more complex mass flow between the first bodies, the third bodies and the environment [16]. Tribocorrosion of stainless steel in high-temperature pressurized water was investigated by Guadalupe et al. [9]. In that study, very thick third bodies consisting of the mixing of metal and metal oxide wear particles were found.

Recently, a comprehensive tribocorrosion model has been proposed for mixed lubricated tribocorrosion situations taking into account the Archard law for mechanical wear (Eq. 3.9 in Chapter 3), the equation for calculating chemical wear (Eq. 4.1)

Fig. 4.5 **a** Plastic deformation at asperity contacts yields: Mechanical wear depassivation and wear-accelerated corrosion. **b** The hydrodynamic flow of the liquid around asperities supports part of the normal load

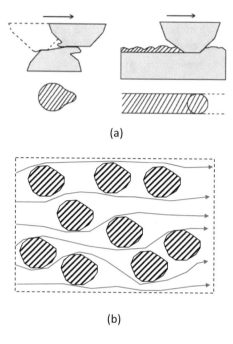

(a)

(b)

and the Hamrock–Dowson's equation for fluid film thickness in lubricated systems [17]. In the model, the normal load F_n is replaced by an effective normal load (F_{eff}) which considers the effect of the hydrodynamic lubrication film on the reduction in load effectively carried out by the contacting asperities (Fig. 4.5).

Equation 4.4 shows the empirical relationship, valid for CoCrMo alloys, between the effective normal load and the hydrodynamic film thickness h_{min} described by the Hamrock–Dowson equation [18].

$$F_{eff} = k_0 \frac{F_n}{(h_{min})^{1.49}} \tag{4.4}$$

where k_0 is a constant of proportionality. The resulting formalism for calculating the total wear in a tribocorrosion system is shown in Eq. 4.5.

$$V_{tot} = k_{mech} \frac{(E')^{0.6556}}{\eta^{0.9685}} \cdot \frac{(F_n)^{1.3129}(v_s)^{0.0315}}{(R')^{1.1473} H}$$
$$+ k_{chem} \frac{M M_r Q_p (E')^{0.3278}}{n F \rho \eta^{0.4843}} \cdot \frac{(F_n)^{0.6565}(v_s)^{0.5158}}{(R')^{0.5737} H^{0.5}} \tag{4.5}$$

where k_{mech} and k_{chem} are constant parameters (to be calibrated for a specific alloy using laboratory tribometers), E' (Pa) is the effective Young's modulus, F_n (N) is the normal force, v_s (m s^{-1}) is the sliding velocity, η (Pa s^{-1}) is viscosity of the solution, R' (m) is the effective radius of curvature, H (Pa) is the surface hardness, M_r (g mol^{-1}) is the molecular mass, Q_p (C m^{-2}) is the passivation charge density, n is the charge number, F is the Faraday's constant, and ρ (g cm^{-3}) is the density of the metal.

References

1. S. Mischler, Sliding tribo-corrosion of passive metals: mechanisms and modeling, in *Tribo-Corrosion Research Testing, Application*, ed. P. Blau, J.P. Celis, D. Drees, F. Friedrich (ASTM international, 2013), pp. 1–18
2. A.H. Zavieh, Tribocorrosion-fatigue (multi-degradation) of stainless steel: A fundamental approach in practical conditions. PhD thesis (2017)
3. Engineering Properties and Service Characteristics, *ASM Specialty Handbook—Copper and Copper Alloys*. ASM International (2001), pp. 385–418
4. D. Drees, J.P. Celis, J.R. Roos, Wear corrosion behaviour of TiN coated tool steel. Matériaux et techniques (1997)
5. D.H. Buckley, Surface Effects in Adhesion, Friction, Wear, and Lubrication. Tribology Series, vol., pp. ii–vi, 1–631 (1981)
6. S. Mischler, S. Debaud, D. Landolt, Wear accelerated corrosion of passive metals in tribocorrosion systems. J. Electrochem. Soc. **145**, 750–758 (1998)
7. A. Bazzoni, S. Mischler, N. Espallargas, Tribocorrosion of pulsed plasma-nitrided CoCrMo implant alloy. Tribol. Lett. **49**, 157–167 (2013)
8. S. Mischler, A. Spiegel, M. Stemp, D. Landolt, Influence of passivity on the tribocorrosion of carbon steel in aqueous solutions. Wear **251**, 1295–1307 (2001). Mischler 1997 wear of engineering materials
9. S. Guadalupe, C. Falcand, W. Chitty, S. Mischler, Tribocorrosion in pressurized high-temperature water: a mass flow model based on the third-body approach. Tribol. Lett. **62**, 10 (2016)
10. S. Cao, S. Mischler, Modelling tribocorrosion of passive metals—a review. Curr. Opin. Solid State Mater. Sci. **22**, 127–141 (2018)
11. H.H. Uhlig, Mechanism of fretting corrosion. J. Appl. Mech. **21**, 401–407 (1954)
12. A.C. Vieira, L.A. Rocha, N. Papageorgiou, S. Mischler, Mechanical and electrochemical deterioration mechanisms in the tribocorrosion of Al alloys in NaCl and in NaNO3 solutions. Corros. Sci. **54**, 26–35 (2012)
13. J. Jiang, M.M. Stack, A. Neville, Modelling the tribo-corrosion interaction in aqueous sliding conditions. Tribol. Int. **35**, 669–679 (2002)
14. J. Jiang, M.M. Stack, Modelling sliding wear: from dry to wet environments. Wear **261**, 954–965 (2006)
15. C.B. Von der Ohe, R. Johnsen, N. Espallargas, Multi-degradation behavior of austenitic and super duplex stainless steel—the effect of 4-point static and cyclic bending applied to a simulated seawater tribocorrosion system. Wear **288**, 39–53 (2012)
16. M. Godet, The third-body approach: a mechanical view of wear. Wear **100**, 437–452 (1984)
17. S. Cao, S. Guadalupe Maldonado, S. Mischler, Tribocorrosion of passive metals in the mixed lubrication regime: theoretical model and application to metal-on-metal artificial hip joints. Wear **324–325**, 55–63 (2015)
18. B. Hamrock, D. Dowson, Elastohydrodynamic lubrication of elliptical contacts for material of low elastic modulus I-fully flooded conjunction. J. Lubr. Technol. **100**(2), 236–245 (1978)

Chapter 5
How to Approach a Tribocorrosion System?

As discussed earlier, tribocorrosion is not simply the sum of wear and corrosion taken individually, but it rather depends on the complex interaction of mechanical and chemical factors. For example, hard PVD TiN coatings are known to provide excellent dry wear resistance and to be chemically inert in aqueous environments; however, they perform very poorly in tribocorrosion situations. The correct design of a tribocorrosion system requires the appropriate selection and combination of materials, geometry and loading and environmental conditions. Due to the complexity of tribocorrosion phenomena, this can be hardly achieved by simplistic empirical approaches but must be based on scientific criteria and on established and robust experimental protocols. Thus, in order to deal with tribocorrosion, the engineer should follow a suitable protocol as the one proposed in Fig. 5.1. The protocol will be described in detail in the next sub-chapter.

5.1 Phase I: Identification of the Problem

A preliminary inspection of the component can provide information about the origin of the damage. The inspection can be carried out simply visually with the help of a magnifying glass or in other cases using advanced equipment such as optical or electron microscopes. Independently on the used technique, in many cases such an inspection is sufficient to identify the problem and finding solutions to it. In other cases, a deeper tribological and corrosion analysis of the system is required.

For example, presence of scratches indicates abrasion (due to insufficient material hardness), presence of material transfer is indicative of adhesive wear (inappropriate material pairing choice or insufficient lubrication), pits could be related to fatigue wear (inappropriate surface finishing or to heavy cyclic loading) or localized corrosion (presence of aggressive ions), extended oxidation in the worn parts can indicate frictional heating, third body build-up, oxidative wear. Unexpected high degradation

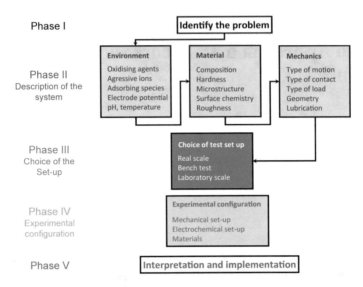

Phase I — Identify the problem

Phase II
Description of the system

Environment	Material	Mechanics
Oxidising agents	Composition	Type of motion
Agressive ions	Hardness	Type of contact
Adsorbing species	Microstructure	Type of load
Electrode potential	Surface chemistry	Geometry
pH, temperature	Roughness	Lubrication

Phase III
Choice of the Set-up

Choice of test set up
Real scale
Bench test
Laboratory scale

Phase IV
Experimental configuration

Experimental configuration
Mechanical set-up
Electrochemical set-up
Materials

Phase V — Interpretation and implementation

Fig. 5.1 Flowchart for dealing with tribocorrosion situation

rates, polished appearance of the contacting surfaces, unexpected surface embrittlement, presence of oxidation products such as rust are typical indicators of a tribocorrosion problem. Case study 1 (Sect. 8.1 in Chapter 8) nicely illustrates how the occurrence of tribocorrosion in nuclear power reactors could be identified by the polished appearance of the worn surfaces.

Figure 5.2 shows an optical image of a cam-follower (40 HRC steel) of an airplane engine. The presence of pits on the surface points out the occurrence of fatigue wear possible due to excessive loading or inappropriate material properties.

Once the tribocorrosion has been identified several solution options arise:

– Act on mechanics by reducing or suppressing the mechanical loading.
– Act on the environment in order to control its corrosivity.
– Select or design appropriate materials.

Fig. 5.2 Damaged cam-follower with pits on the bearing surface (arrow indicated)

Common to these approaches is a pertinent description of the environmental, material and mechanical operating conditions, as illustrated in the following section.

5.2 Phase II: Description of the System

The description of the system consists of obtaining detailed quantitative information as much as possible from three main aspects: environment, material and contact mechanics. This is a prerequisite for identifying possible material degradation mechanisms, defining laboratory tests, material specifications, mechanical redesign of components or corrosion suppression.

5.2.1 Environment

The first key point is to gather as much complete information as possible about the environment. The most relevant issue is to identify the oxidizing agents. Often, a simple measure of the pH can give useful information: at pHs lower than 4 the most active oxidizing agent is protons whereas at neutral or basic pHs the active oxidizing agent is molecular oxygen dissolved in water (see Chap. 2). Measurement of the OCP can be a practical alternative to identify the oxidative characteristics of the environment with respect to the material (see Chap. 2). Note that under experimental laboratory conditions it is easier to reproduce the OCP than the often-complex chemistry found in engineering applications. The presence of aggressive ions such as chlorides and other halogens should also be checked as they impede or slow down repassivation and can cause localized corrosion (i.e. pitting leading possibly to fatigue crack initiation). Adsorbed species such as organic molecules, phosphates or other species act on the surface charge of the worn particles determining the agglomeration of debris and may therefore interfere in the build-up of passive films. Figure 5.3 shows an example how the chemistry of the environment, specifically the presence of adsorbed species such as the protein albumin, influences the tribocorrosion material loss of a CoCrMo alloy sliding against an alumina ball [1]. Depending on heat treatment the influence of albumin promotes or reduces material loss. This effect was related to the influence of the surface adsorbed species on the debris behaviour.

In case study 8.3 (Sect. 8.3 in Chapter 8), the measurement of Evans diagrams of the punched steel in the tested lubricants revealed that the presence of passivating inhibitors rendered this material highly sensitive to wear-accelerated corrosion.

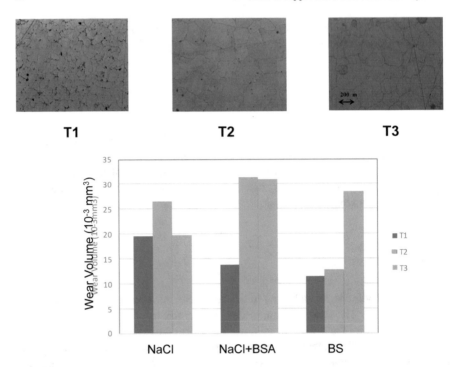

Fig. 5.3 Microstructures of a high-carbon CoCrMo alloy after different thermal treatments, T1-T2-T3 (left) and wear volumes of the materials after tribocorrosion tests in different solutions (NaCl, NaCl + bovine serum albumin (BSA) and bovine serum (BS)). From [1]

5.2.2 Material

Defining the materials in contact is another essential parameter in tribocorrosion. Besides the composition of the materials in contact, their microstructure plays an important role in tribocorrosion. The example of Fig. 5.3 also shows how the microstructure of a CoCrMo alloy that was modified by heat treatments changed the total material loss under tribocorrosion conditions [1]. Surface composition and thus properties can differ from that of the bulk. Heat treatments experienced by, or performed at, the component can generate gradual microstructural changes from the surface to the bulk. This can lead to differences between the hardness of the surface and that from the bulk. Similar differences can be introduced during machining or sample preparation.

5.2.3 Mechanics

Definition of the mechanics of the system includes not only the geometry of the contact but also the lubrication conditions. The type of contact and motion (Fig. 5.4) determine the mechanical conditions of the system. For example, depending on whether the contact is conformal or non-conformal, the amplitude and distribution of the mechanical stresses vary.

The load can be applied continuously or cyclically with unidirectional or reciprocating motion (Fig. 5.5).

The geometry of the samples (e.g. ball-on-flat, cylinder-on-cylinder, block-on-ring, pin-on-disk) and the size may mimic different tribological situations but also modify the heat dissipation during the contact.

Finally, the lubricating regime (boundary, mixed and/or hydrodynamic) should be defined according to the type of lubricant used (viscosity), the type of motion (speed) and the load applied.

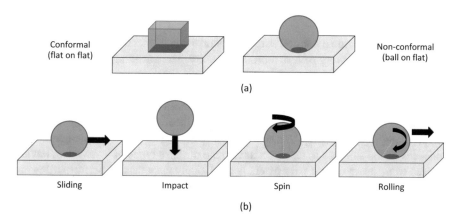

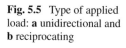

Fig. 5.4 a Type of contacts and **b** type of motion in tribological experiments

Fig. 5.5 Type of applied load: **a** unidirectional and **b** reciprocating

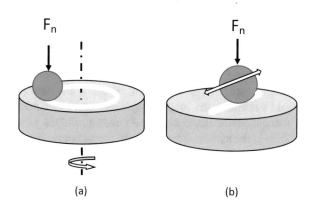

5.3 Phase III: Choice of Experimental Set-Up

There are three possibilities to experimentally assess tribocorrosion, namely real-scale tests, bench tests and laboratory tests. The choice depends on the goal that one wants to achieve. When the goal is to reproduce the real situation and to know the response of the system or component under conditions identical to the operating ones, real-scale tests or bench tests should be the choice. When the aim is to understand the degradation mechanisms of a material by quantifying wear, measuring friction and changing the lubricating regime, then laboratory tests allow for an idealized situation in which all parameters of the system (environment, material and mechanics) can be properly defined.

In real-scale tests, components subjected to tribocorrosion are directly tested in the real engineering system during operation. The advantage is that there is no need for a detailed analysis of the system since it is typically well-known and defined; however, the disadvantages are usually long response times, undefined test conditions rendering any difficult extrapolation of results to other situations, relatively high costs and the impossibility in many cases to set up measuring devices for friction or other output parameters.

Bench tests provide a method to test tribocorrosion of mechanical components in test conditions that more or less reproduce reality. The advantages cover the use of the real components with the same surface features, better controlled test conditions than real-scale tests and the possibility to accelerate tests and to measure output parameters such as friction, displacement, temperature or electrode potential.

Laboratory tests are usually carried out under idealized, but well-controlled and defined conditions allowing for a short response time. The use of simple geometries and sizes of the samples to test is an advantage of these types of tests. Also, having devices to measure output parameters is a relatively easy task to perform allowing to gather as much information as possible from the process taking place during the test. A laboratory tribo-electrochemical measurement revealed in a very short time the sensitivity of steel in oil/water emulsions to wear-accelerated corrosion and permitted identify possible ways to minimize the damage of metallic tools used in drawn and wall ironing for aluminium cans. However, scaling up of the results can be challenging unless reliable theoretical models are available.

Figure 5.6 shows an example of three different experimental configurations for biomedical applications. The goal is to characterize and determine the tribocorrosion response of hip joint implant materials. The system can only be tested under real conditions by monitoring the state of the implant in a patient. Of course, in this specific case, the operating conditions are just partially known (as an example one could determine the loading, dynamic and kinematics, by a motion study) but the chemical and/or electrochemical conditions of the joint can be hardly determined. In addition, the system (human body) cannot provide output information such as friction. Therefore, this case of real test can only be carried out by analysis of the loading conditions, in situ visualization of the implant and surrounding tissues through NMR

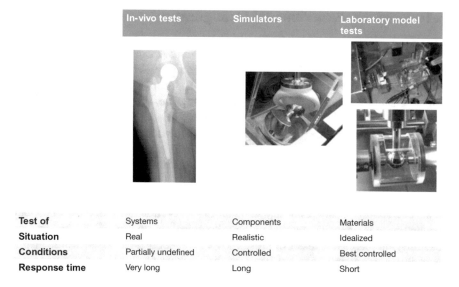

Test of	Systems	Components	Materials
Situation	Real	Realistic	Idealized
Conditions	Partially undefined	Controlled	Best controlled
Response time	Very long	Long	Short

Fig. 5.6 Example of different experimental set-ups for studying tribocorrosion of materials used in biomedical hip joints (image of in vivo tests from http://www.hipandkneeadvice.com)

techniques and by analysing the implant once extracted from the patient. The study of explants would allow identifying the cause of failure and the main degradation mechanism occurring in the system after it fails.

Biomedical implants can also be tested using instruments such as hip simulators. In this case, it is possible to reproduce the operating conditions of the hip prosthesis by testing the real components, but in simulated body fluids. In this case, the geometry, the material and the loading conditions can be reproduced but not the chemical composition of the environment, neither the mechanical interaction between the body elements (bones, tissues) with the implant itself. These tests allow for checking the behaviour of the components used in the real geometry of the hip joints and obtaining actual output data, although the limitation of the unknown properties of the environment makes the obtained output difficult to extrapolate to the real practice.

Finally, the materials used in the fabrication of hip implants can be tested in laboratory tribometers, which allow an almost perfect control of the operating conditions in an idealized way. With the output obtained from these tests, the material behaviour can be defined under a relatively high variety of conditions (chemical, electrochemical and mechanical) and also theoretical models allowing for scale up can be developed (as clearly shown in Cao et al. [17] work, Chap. 4).

Finally, the selection of the type of test depends on several factors including the availability of the equipment and cost.

5.4 Phase IV: Experimental Configuration

Although all tribological tests can be combined with electrochemical measurements, only the ones specifically applied to tribocorrosion studies will be described in this section. The electrochemical techniques used in tribocorrosion studies are based on a three-electrode set-up composed of the working (i.e. the material under investigation), the reference and the counter (or auxiliary) electrodes. The reference electrode serves to measure the electrode potential of the working electrode. The current flowing between the working and the counter electrode at an applied potential through a potentiostat corresponds under specific conditions to the corrosion rate. These techniques are also widely used in corrosion studies and can be extended to tribocorrosion situations in order to understand the role of electrochemical and chemical reactions in the degradation mechanisms and to quantify the reaction rates. In order to do that, a potentiostat is typically coupled to the tribological set-up (i.e. tribometer, fretting machine, microabrasion tester or erosion rig). A potentiostat is an electronic instrument that imposes the potential between the working electrode and the reference electrode by adjusting the current between the working and the counter electrodes. In some cases, the simple measurement of the potential difference between the working electrode and the reference electrode, OCP provides enough information about the tribocorrosion system; thus, in that case, only a voltmeter can be used. A more detailed description of how to carry out tribocorrosion tests is given in Chap. 6. Figure 5.7 shows an example of a reciprocating sliding tribometer coupled to a potentiostat and a three-electrode electrochemical cell (electrochemical tribometer).

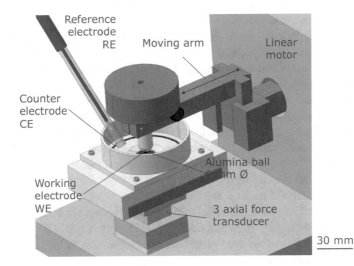

Fig. 5.7 Electrochemical tribometer for tribocorrosion testing

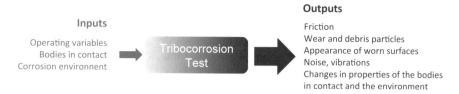

Fig. 5.8 A tribocorrosion test can be seen as a process in which certain inputs are converted into measurable outputs

Although a wide variety of electrochemical techniques is available for corrosion studies [2], only the most commonly used in tribocorrosion studies will be described in Chap. 6 (open circuit potential measurements, potentiostatic and potentiodynamic tests).

Other analytical measurements can be coupled with the electrochemical techniques to complement the tribocorrosion tests, for example measuring the metal ion release into the electrolyte by inductively coupled plasma (ICP) [3] or measuring the pH. These techniques, however, are beyond the scope of this book and will not be further discussed.

A tribocorrosion test can be therefore described as a process in which specific inputs generate a set of outputs for the monitoring, quantification and analysis of the tribocorrosion phenomenon. The scheme in Fig. 5.8 shows the different set of inputs in a tribocorrosion test and the generated outputs. In Table 5.1 the input parameters are defined.

The reader is referred to the round-robin study on tribocorrosion tests by Mischler and Ponthiaux [4] to have reference measurements and to "Part II Method for measurement and prevention of tribocorrosion in the Tribocorrosion of passive metal

Table 5.1 Input parameters of a tribocorrosion system	Operating variables	Type of motion Duration of operation Load Velocity Temperature Electrochemical conditions
	Bodies in contact	Material Geometry Dimensions Chemical composition Hardness Topography
	Corrosion environment	Viscosity Chemistry pH Ion concentration

and coatings" book and to "Tribo-Corrosion: Research, Testing, and Applications" for more detailed protocols in tribocorrosion testing [5, 6].

Finally, depending on the tribological situation designed in the experimental set-up, one can distinguish between sliding, fretting, microabrasion or erosion tribocorrosion tests. These will be described in Chap. 6.

5.5 Phase V: Interpretation and Outcomes

As shown in Fig. 5.8, the output of a tribocorrosion test can be electrochemical (E, i), mechanical (F_t, V), material (H, surface chemistry) and chemical (pH, ion concentration) parameters. These outputs can be used to quantify the origin (mechanical or chemical) of the material loss using models. In addition, it can help interpreting the involved phenomena.

For example, high contribution of the mechanical part of the total material loss in a tribocorrosion system can be tackled by modifying the mechanical aspects of the system (i.e. by reducing the loads or modifying surface roughness). On the other hand, a high contribution chemical part of the total material loss should be approached by considering not only the mechanical aspects but also the chemical and electrochemical conditions of the system (i.e. addition of corrosion inhibitors or modification of electrode potential). In the case studies, Chap. 8, dedicated examples on how to interpret tribocorrosion outcomes will be given.

References

1. L. Casaban Julian, Igual Munoz A, Influence of microstructure of HC CoCrMo biomedical alloys on the corrosion and wear behaviour in simulated body fluids. Tribol. Int. **44**, 318–329 (2011)
2. D. Landolt, *Corrosion and Surface Chemistry of Metals* (EPFL Press, Lausanne, 2007)
3. N. Espallargas, C. Torres, Muñoz A. Igual, A metal ion release study of CoCrMo exposed to corrosion and tribocorrosion conditions in simulated body fluids. Wear **332–333**, 669–678 (2015)
4. S. Mischler, P. Ponthiaux, A round robin on combined electrochemical and friction tests on alumina/stainless steel contacts in sulphuric acid. Wear **28**, 211–225 (2001)
5. D. Landolt, S. Mischler, *Tribocorrosion of Passive Metals and Coatings* (Woodhead Publishing, Lausanne, 2011)
6. P. Blau, J.P. Celis, D. Drees, F. Franek (eds.), *Tribo-Corrosion: Research, Testing, and Applications, STP1563-EB* (ASTM International, West Conshohocken, PA, 2013)

Chapter 6
Experimental Techniques for Tribocorrosion

6.1 Sliding Tribocorrosion

The most common tribometers used for studying sliding tribological contacts consist of a counterpart (typically a ceramic ball) moving linearly (alternating back and forth) or unidirectionally (circular trajectory) on a metallic surface (working electrode). Typically, two electrochemical set-ups are combined with those tribometers, the simple measurement of OCP and the potentiostatic set-up where the potential is imposed to mimic specific corrosion conditions. Figure 6.1a shows the two most common configurations used for commercial tribometers and Fig. 6.1b the corresponding tribocorrosion set-up when the electrode potential is measured or imposed simultaneously to the sliding.

In the open circuit potential technique when rubbing starts, a potential decay is observed as a consequence of the mechanical removal of the passive layer, thus causing an acceleration of metal dissolution in the rubbed area [1]. A typical OCP evolution with time during the open circuit potential technique for a passive metal is shown in Fig. 6.2. This technique shows whether and to which extent a galvanic couple between the worn area (anode) and the rest of the electrode surface (cathode) is established. The OCP decay registered during rubbing can be related to the material loss caused by the enhanced corrosion kinetics due to wear according to a theoretical model proposed by Vieira et al. [1]. In Fig. 6.2 the OCP recovers its initial value after rubbing stops at a specific rate. However, it is possible to find cases in which the potential does not fully recover the initial value due to the permanent damage created on the metal surface.

Registering the OCP in a tribocorrosion test gives an in situ measure of the degradation level of the material. Figure 6.3 [2] shows the OCP evolution with time before, during and after sliding of a stainless steel, uncoated and coated with CrSiN and DLC coatings. When the uncoated material, 316L stainless steel, is subjected to tribocorrosion conditions, its OCP abruptly decreases at the onset of rubbing. The OCP evolution with time of the coated materials is different, in the CrSiN coating a progressive decrease in the OCP was observed while the OCP values did not change

© The Author(s), under exclusive license to Springer Nature Switzerland AG 2020
A. Igual Munoz et al., *Tribocorrosion*, SpringerBriefs in Applied
Sciences and Technology, https://doi.org/10.1007/978-3-030-48107-0_6

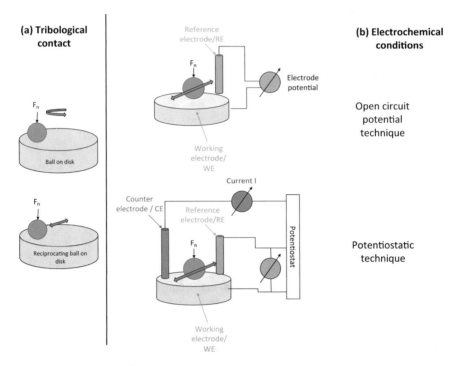

Fig. 6.1 Laboratory experimental set-up for tribocorrosion tests, showing two examples of tribological contacts (**a**) and electrochemical configurations (**b**)

Fig. 6.2 Time evolution of the open circuit potential in a tribocorrosion test of an aluminium alloy in NaCl (from [1])

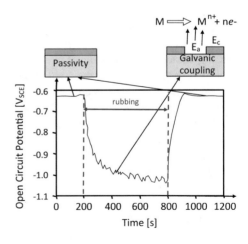

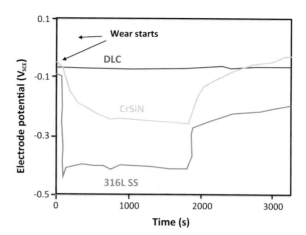

Fig. 6.3 Time evolution of the open circuit potential of a stainless steel and the same stainless steel coated with CrSiN and DLC (from [2])

in the case of the DLC coating. This figure clearly shows that some materials do not change their electrochemical nature during rubbing (i.e. DLC) while others do.

Another example of coating characterization through OCP tribocorrosion tests is illustrated in detail in case study 5 (Sect. 8.5 in Chapter 8).

Using the potentiostatic technique, the potential is fixed and the current evolution with time is measured. The current evolution with time depends on the material–electrolyte combination as well as on the applied potential, normal load and sliding velocity. Figure 6.4 shows, for example, a clear increase in current at the onset of rubbing due to the mechanical depassivation of the rubbed surface. This clearly indicates an increase in corrosion kinetics due to the mechanical damage applied onto the surface.

The current evolution with time during tribocorrosion tests can be used to determine the chemical wear volume, that is, the metal volume released due to the mechanical activation by the cyclic depassivation/repassivation process during rubbing.

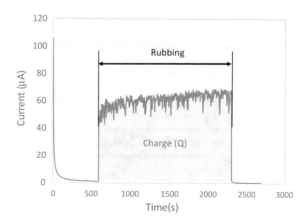

Fig. 6.4 Typical current evolution with time in a tribocorrosion test carried out under potentiostatic conditions. The blue area corresponds to the integrated excess current recorded during rubbing

From the current, the charge can be determined by integrating the excess current recorded during rubbing after subtracting the background current (average current from unworn areas before and after rubbing). Figure 6.4 illustrates this procedure where the charge Q is integrated as the area between the start and finish of rubbing. The chemical volume can be thus calculated using Faraday's law (Eq. 6.1) once the charge is known from the experiment.

$$V_{\text{chem}} = \frac{Q \cdot M_r \cdot}{\rho \cdot n \cdot F}$$ (6.1)

where Q is the charge flowing from the wear track during the sliding period (C), M_r is the atomic mass of the metal (g/mol), ρ is the density of the metal (g/cm), F is the Faraday's constant (95485 C/mol), and n is the oxidation valence.

Another contribution to the total material loss in tribocorrosion from the mechanistic approach is the mechanical wear term. The mechanical term (V_{mech}) is related to the material removed as metal particles by the mechanical action of the counter material during the sliding process. In tribocorrosion this is determined by subtracting the V_{chem} from the total wear (V_{tot}, i.e. the volume of the wear track created after rubbing).

Depending on the prevailing electrochemical conditions (i.e. applied potential) different currents are measured, and therefore different material loss is generated, but also different mechanical wear can be observed. Figure 6.5 shows an example of the mechanical and chemical contributions to the total wear volume of CoCrMo alloy immersed in a sulphuric acid solution rubbing against an alumina ball [3]. At cathodic potentials (below the corrosion potential), the total wear volume is low and only caused by the mechanical action of the tribological contact (no corrosion occurs). At passive potentials (above the corrosion potential) when a thin oxide film is present on the surface of the metal, the total wear volume is higher and the chemical

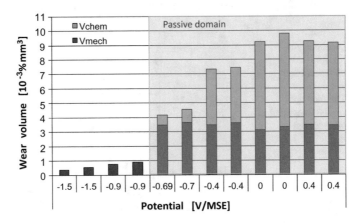

Fig. 6.5 Chemical and mechanical wear volume of CoCrMo alloy tested at different applied potentials in H_2SO_4 (from [3])

contribution appears as a consequence of the mechanical removal of the passive film and the subsequent wear-accelerated corrosion, but the mechanical wear damage also increases. Therefore, the presence of a thin oxide film modifies the mechanical response of the metal under tribocorrosion conditions.

6.2   Fretting-Corrosion

When the tribological experiment carried out in an aqueous electrolyte under controlled electrochemical conditions is produced by a small amplitude oscillatory motion (often originated by vibrations) between two solid surfaces in contact, the studied phenomenon is referred to as fretting-corrosion. Typical examples of situations in which fretting may occur are in riveted or bolted joints, between strands of wire ropes or between the rolling elements and their tracks in the bearings. Essentially, sliding and fretting corrosion tests are very similar with only some particularities of the fretting situation to be considered:

- Due to the small amplitude of the motion, a significant part of the displacement is accommodated by elastic deformation. During elastic accommodation no depassivation takes place (Fig. 6.6) [4].
- A large part of the rubbing surface is in permanent contact with the counterpart material. As a consequence, debris tends to remain trapped within the contact and therefore changes in the chemistry of the electrolyte solution can occur (i.e.

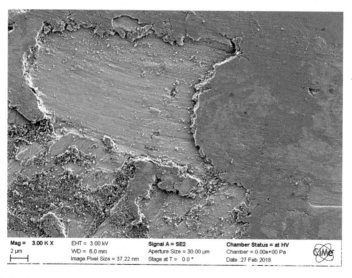

Fig. 6.6 SEM image of a Ti6Al4V wear track tested under fretting-corrosion conditions in NaCl 0.9%wt at 0.3 V_{SHE}, normal load 10 N, displacement 50 μm and frequency 1 Hz. From [5]

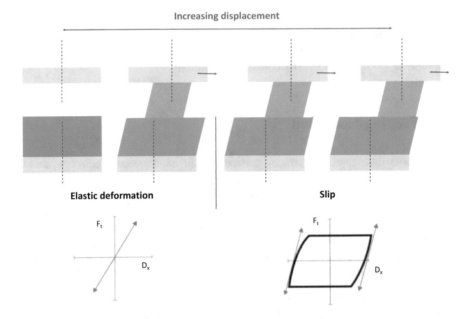

Fig. 6.7 Illustration of fretting regimes and fretting loops

acidification, oxygen depletion, etc.). The trapped particles are exposed simultaneously to wear, corrosion and deformation typically generating a third body. Figure 6.6 shows the third body generated on the surface of a Ti6Al4V alloy together with some fragmented wear debris which could be either ejected from the contact or be re-incorporated to the third body.

Figure 6.7 illustrates the different fretting situations depending on the displacement and the corresponding fretting loops (a representation of the tangential force F_t versus amplitude displacement, D_x during one oscillatory cycle). Monitoring the different fretting regimes by measuring the evolution of the force and the displacement with time allows identifying the elastic and plastic accommodation of the system. Each regime typically generates a mode of surface damage. At low displacements, mainly elastic deformation takes place, thus the tangential force equals the normal load times the coefficient of friction. Stresses produced by the small displacement are accommodated by the elastic deformation in the contact and low or negligible wear occurs. At higher displacements, the tangential force is higher than the normal load multiplied by the coefficient of friction and the slip at the interface accommodates plastically the stresses in the contact. The plateau in the tangential force indicates that the displacement imposed to the contact results in relative sliding between the two counter bodies. In this situation, higher material damage occurs due to the combined effect of wear and corrosion.

Barril et al. [4] studied the influence of fretting regimes on the tribocorrosion behaviour of Ti6Al4V rubbed against an alumina ball in 0.9%wt NaCl and recorded

Fig. 6.8 Transients of
a lateral position,
b tangential force and
c current of Ti6Al4V rubbed
against an alumina ball at
100 μm of displacement
amplitude in 0.9%wt NaCl.
From [4]

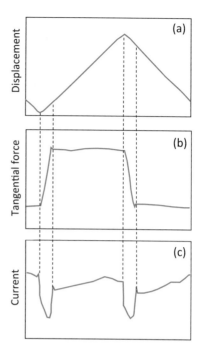

the current response of the titanium alloy at an imposed passive potential during fretting-corrosion at high amplitude displacements (Fig. 6.8). Figure 6.8 shows an example of one of the transients of the lateral position of the counterpart (Fig. 6.8a), the measured tangential force (Fig. 6.8b) and the current response (Fig. 6.8c) of a Ti6Al4V rubbed against an alumina ball. In this work, depassivation occurred during sliding and stopped during the elastic accommodation of the imposed displacement. Therefore, the fretting regime plays a crucial role for the electrochemical response of the system, producing a material damage by wear, corrosion and wear-accelerated corrosion under slip regime. Clearly, the depassivation/repassivation phenomenon occurs while sliding, thus wear-accelerated corrosion mechanism contributes to the total wear damage of the system. On the contrary, when the displacement was fully accommodated in the contact, no wear-accelerated corrosion was observed.

6.3 Microabrasion Corrosion

Microabrasion corrosion is a special tribocorrosion situation where a body, usually a ball, slides against a counterpart in the presence of a slurry. The slurry causes third body abrasion and enhances wear and corrosion leading to a considerably shortage of the service life of components. The microabrasion ball-cratering test rig was specifically developed for tribocorrosion. It basically consists of a sample (working electrode in a three-electrode electrochemical cell) loaded against a ball, which rotates parallel to the plane of the sample, while the abrasive slurry is drop-fed onto the ball at a certain rate. Figure 6.9a shows the experimental set-up used by Sun et al. [6] to study the abrasion–corrosion behaviour of CoCrMo alloy in simulated body fluids. They measured the current response of the CoCrMo in a potentiostatic test in the electrolyte containing SiC particles (Fig. 6.9b)

Figure 6.10 shows an example of the current evolution with time under potentio-static conditions of a 2205 duplex stainless steel in a microabrasion–corrosion test carried out in 3.5% NaCl with different SiC particles (volume concentration). A sharp

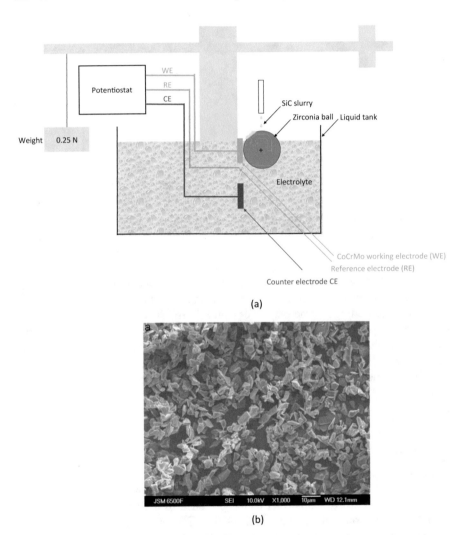

(a)

(b)

Fig. 6.9 Scheme of the microabrasion corrosion set-up (**a**) and SEM image of the SiC abrasives (**b**) [6]

Fig. 6.10 Influence of
particle concentration on
current during
abrasion–corrosion tests.
From [7]

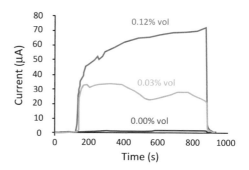

increase in the current is observed at the onset of the microabrasion–corrosion test
due to the wear-accelerated corrosion by cyclic depassivation/repassivation mechanisms. The increase in particle concentration in the solution increases the current,
thus abrasion accelerates the corrosion rate of the material.

In the same manner as that in a sliding tribocorrosion tests, the material loss due
to the electrochemical dissolution can be calculated using the Faraday's law. Then,
the wear performance of the material can be studied depending on different testing
parameters (i.e. particle size, number of particles) by analysing the dependency of the
specific wear rate (mm^3 N^{-1} m^{-1}) and the current during abrasion–corrosion. This
dependency can then be exploited to improve the understanding of tribocorrosion
systems and their optimization.

6.4 Erosion–Corrosion

Wear caused by impingement of particles, fluids or gases is named erosion. When
erosion and corrosion simultaneously take place, the phenomenon is called erosion–
corrosion (typically encountered in pumps and pipes exposed to turbulent flow in the
presence of suspended particles). Although erosion–corrosion may take place in a
liquid or gas (slurry erosion or high-temperature erosion–corrosion, respectively), in
this section only experimental techniques for studying the slurry erosion–corrosion
phenomenon will be described. Figure 6.11 shows two typical erosion–corrosion rigs
for slurry (Fig. 6.11a), and for jet impingement (Fig. 6.11b).

The slurry pot erosion test rig (Fig. 6.11a) consists of a rotating cylinder electrode with control velocity immersed in an electrolyte containing solid particles and
coupled with a potentiostat to measure the electrochemical signal simultaneously for
the erosion effect. The working electrodes are placed in the extremes of a rotating
arm immersed in the electrolyte which contains the solid particles. The rotation of
the arms allows to control the linear speed of the slurry at the specimen surface with
a specific erosion attacking angle.

The electrochemical tests used in this experimental set-up are potentiodynamic
tests carried out under stagnant and under erosion–corrosion conditions [10, 11]

(a)

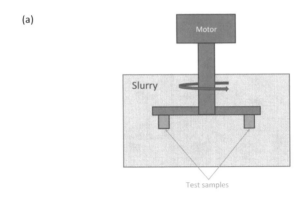

(b)

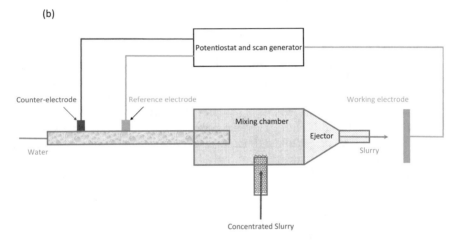

Fig. 6.11 **a** Slurry pot erosion tests rig from [8], **b** Jet impingement slurry erosion corrosion rigs [9]

and potentiostatic tests at imposed passive potential are carried out to quantify the corrosion enhancement due to erosion [10].

In the jet impingement rig (Fig. 6.11b), the aqueous phase is pumped into an ejector nozzle assembly, creating a differential pressure that helps entering the concentrated slurry from a slurry reservoir. In the central chamber the solid particles are mixed with the electrolyte and then the diluted slurry is formed within the ejector and it is exited at the ejector nozzle to impact the material at a certain velocity with a specific angle (impact angle). The material is connected to the working electrode terminal of a potentiostat; thus the potential and/or current is measured during the test. The reference and counter electrodes can be placed at the entrance (as shown in the figure) or at the exit of the injector chamber.

Figure 6.12 shows an example of the potentiodynamic curves recorded for a mild steel in a carbonate/bicarbonate solution carried out in a slurry pot erosion–corrosion apparatus at different velocity flows in the presence of alumina particles. It is possible

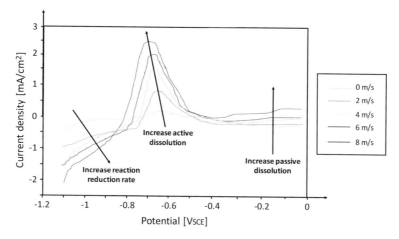

Fig. 6.12 Potentiodynamic curves of a mild steel in 0.5 M NaHCO$_3$ + 0.5 M Na$_2$CO$_3$ containing 300 g/L of alumina particles at different velocities. From [10]

to observe how the active dissolution of the steel increases with velocity. The passive current density also increases with the velocity, thus suggesting that impinging particles cause local disruption of the passive state initiating a depassivation/repassivation processes.

This trend of increase in passive dissolution with the velocity was also confirmed in potentiostatic tests. In Fig. 6.13 one can observe that the increase in velocity leads to an increase in the current in the erosion–corrosion tests due to the increase in the kinetic energy of the particles. This is due to the mechanical rupture of the passive film and the deformation of the underlying material surface. The subsequent increase in the frequency of the particle impact with the increase in the flow velocity also causes an increase in the material removal.

Fig. 6.13 Current trend during erosion–corrosion of AISI316L stainless steel tested in a slurry erosion rig with silica particles used as erodent in 3.5% NaCl electrolyte. From [12]

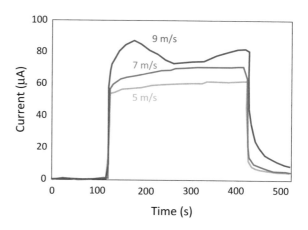

References

1. A.C. Vieira, L.A. Rocha, N. Papageorgiou, S. Mischler, Mechanical and electrochemical deterioration mechanisms in the tribocorrosion of Al alloys in NaCl and in NaNO3 solutions. Corros. Sci. **54**, 26–35 (2012)
2. D. Landolt, S. Mischler, *Tribocorrosion of Passive Metals and Coatings* (Woodhead Publishing, Lausanne, 2011)
3. S. Guadalupe Maldonado, S. Mischler, M. Cantoni, W.J. Chitty, D. Hertz, Mechanical and chemical mechanisms in the tribocorrosion of a stellite type alloy. Wear **308**, 213–221 (2013)
4. S. Barril, S. Mischler, D. Landolt, Influence of fretting regimes on the tribocorrosion behaviour of Ti6Al4V in 0.9%wt sodium chloride solution. Wear **256**, 963–972 (2004)
5. A. Bermudez Castaneda, Degradation of modular hip joint implants a corrosion and tribocorrosion approach. Ph.D. (2018)
6. D. Sun, J.A. Wharton, R.J.K. Wood, Abrasive size and concentration effects on the tribocorrosion of cast CoCrMo alloy in simulated body fluids. Tribol. Int. **42**, 1595–1604 (2009)
7. R.J.K. Wood, D. Sun, M.R. Thakarea, A. de Frutos Rozas, J.A. Wharton, Interpretation of electrochemical measurements made during micro-scale abrasion-corrosion. Tribol. Int. **43**, 1218–1227 (2018)
8. S.S. Rajahram, T.J. Harvey, R.J.K. Wood, Erosion–corrosion resistance of engineering materials in various test conditions. Wear **267**, 244–254 (2009)
9. R.C. Barik, J.A. Wharton, R.J.K. Wood, K.R. Stokes, Electro-mechanical interactions during erosion–corrosion. Wear **267**, 1900–1908 (2009)
10. M.M. Stack, S. Zhou, R.C. Newmann, Identification of transitions in erosion-corrosion regimes in aqueous environments. Wear **186–187**, 523–532. (1995, 2004)
11. M. Jones, R.J. Llewellyn, Erosion–corrosion assessment of materials for use in the resources industry. Wear **267**, 2003–2009 (2009)
12. S.S. Rajahram, T.J. Harvey, R.J.K. Wood, Electrochemical investigation of erosion–corrosion using a slurry pot erosion tester. Tribol. Int. **44**, 232–240 (2011)

Chapter 7
Characterization of Worn Surfaces

Surface topography, surface chemistry and microstructure below the rubbing surface may change as a consequence of the tribocorrosion phenomena. Therefore, a detailed observation and analysis of the worn surfaces gives qualitative and quantitative information about prevailing tribocorrosion mechanisms and causes of failure.

7.1 Morphology of Worn Surfaces

In a first simple approach, worn surfaces can be observed by optical microscopy in order to identify patterns characteristic of specific mechanisms (e.g. scratches are typical of an abrasive phenomenon). Figure 7.1 shows an example of optical images of worn areas of ruthenium in 0.01 M H_2SO_4 solution at different applied potentials. At the lowest potential (Fig. 7.1a) the image shows a very smooth surface with only isolated grooves and some slip bands at the edge of the wear track. At higher applied potentials (Fig. 7.1b) the wear track is wider and more scratches appear.

For a more detailed analysis of those patterns, scanning electron microscopy (SEM) can be used. Scanning electron microscopy (SEM) is a very widespread method for the characterization of surface structure and topography of materials in a non-destructive mode. This technique is well suited for the study of micrometre-scale or even millimetre-scale surface topography (i.e. observation of wear debris). In Fig. 7.2 the scratches observed at higher applied potentials, which are produced by the abrasive metallic particles generated during the test, are also clearly visible in the SEM images (Fig. 7.2b). From this image, it appears that the size of the wear particle corresponds to the scratch width in the ruthenium disk. Moreover, the SEM images show more clearly the ridges and that they are formed by smeared material (Fig. 7.2a).

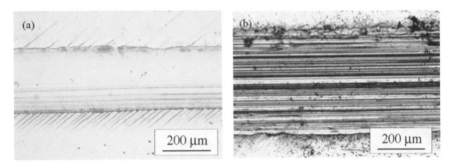

Fig. 7.1 Optical microscopy images of ruthenium worn areas at two different applied potentials, **a** -1 V_{MSE} and **b** -0.25 V_{MSE}. From [1]

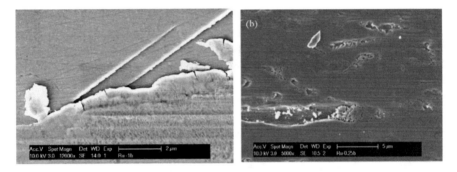

Fig. 7.2 SEM images of ruthenium worn areas at two different applied potentials, **a** -1 V_{MSE} in the edge of the wear track and **b** -0.25 V_{MSE} in the middle of the wear track. From [1]

7.2 Subsurface Transformations

Metallurgical transformations are also known to occur during tribocorrosion conditions in the subsurface caused by multiaxial fatigue stress due to the cyclic friction effects. These phenomena are the origin of the plastic strain accumulation, the subsurface deformation and the subsequent microstructural transformations (i.e. nanocrystalline layer formation) which in turn may influence the wear behaviour of sliding contacts. Indeed, a layered structure from a very deformed nanocrystalline zone, a strain-hardened layer towards the bulk materials was identified by several authors [2, 3]. These subsurface structures may affect mechanical wear by different mechanisms (i.e. rotation and detachment of nano-grains forming nanometric wear particles [4]. Several techniques are appropriate to characterize those subsurface structures and chemical composition. Among them, focused ion beam (FIB) system is a recently introduced powerful technique. It operates as a scanning electron microscope (SEM) and uses a finely focused beam of gallium ions for milling the material. The ion beam allows for precisely cutting a cross-section of the sample surface.

Fig. 7.3 FIB cross-section perpendicular to the wear track of an asutenitic stainless steel, ASS, rubbed against an alumina ball at 70 N normal loads. From [5]

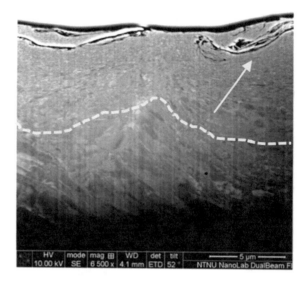

Figure 7.3 shows the focus ion beam (FIB) image of the worn zone of a stainless steel after tribocorrosion test in NaCl solution. The layered structure shows a grain refinement caused by the severe plastic deformation which could be divided in different zones. The first top layer constituted by a finely recrystallized material and the second one with deformed grains. In the image, the formation of cracks promoted by the high load is also observed surrounded by severely deformed and recrystallized grains. These cracks could be the source of wear debris formation.

7.3 Surface Chemistry Modification

Material response under tribocorrosion conditions can also be influenced by the chemical composition of surfaces which may differ from those of the bulk material due to the effect of different phenomena (i.e. adsorption, oxidation). It has been well documented for example that the presence of a thin oxide layer increases the mechanical wear of most common engineering alloys (such as stainless steel [5, 2]), CoCr [6], ruthenium [1] or NiCr [7]. Indeed, several experimental results show the crucial role of surface oxidation on wear-accelerated corrosion and on surface mechanical phenomena (i.e. Roscoe effect, see Chap. 4). From a chemical point of view, surface chemistry defines the intervening layers in a contact, determining its chemical compatibility, shear strength and lubricant properties (boundary layers).

Surface analysis methods refer to those experimental techniques that give chemical information on surfaces to a depth of 2–3 nm or less. These methods are able to measure concentration variations over a depth of tens or even hundred of nanometres

Table 7.1 Characteristics of the most common surface analysis methods [8]

	XPS	AES	SIMS
Incident radiation (energy in KeV)	Photons (1–2)	Electrons (1–10)	Ions (0.5–2)
Analysed radiation	Photoelectrons	Auger electrons	Scattered ions
Analysed depth (atomic layers)	3–15	2–10	2–10
Detection limit (% atomic layers)	0.1–2	0.1–1	10^{-4}–1

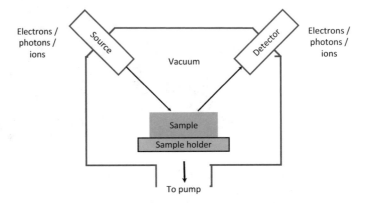

Fig. 7.4 Illustration of the operating system of different surface analysis techniques

when combined with ion sputtering depth profiling by using a similar ion beam as the FIB. The most common methods are listed in Table 7.1.

The principle of operation in all cases is illustrated in Fig. 7.4. It consists of the irradiation of the surface with electrons, X-rays or ions and the detection (quantity and energy or mass) of the emitted electrons, photons or ions, respectively, as a result of collisions of the incident particles with atoms of the solid. The emitted species contain specific chemical information. These techniques require an ultra-high vacuum chamber.

A detailed analysis of the chemical composition of the wear track by carrying out Auger electron spectroscopy (AES) provides information about the modification of the surface layers resulting from the simultaneous action of wear and corrosion. Figure 7.5 shows two AES profiles: outside (Fig. 7.5a) and inside (Fig. 7.5b) the wear track after tribocorrosion tests at high temperature. The graphs show a thicker oxide film inside the wear track compared to the initial passive layer of the stainless steel (outside) formed by mixed oxide products indicating the presence of a third body through which wear occurs.

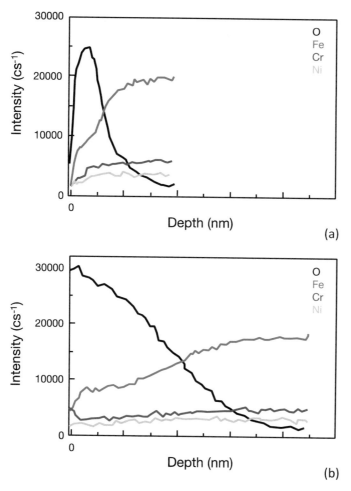

Fig. 7.5 AES profiles, **a** outside and **b** inside the wear track of an AISI 304L after a tribocorrosion test carried out in water at pH 6.9 (300 °C and 154 bars) and OCP during 70 h at 10 Hz and a normal load of 10 N. From [9]

References

1. J. Stojadinovic, L. Mendia, D. Bouvet, M. Declercq, S. Mischler, Electrochemically controlled wear transitions in the tribocorrosion of ruthenium. Wear **267**, 186–194 (2009)
2. J. Perret, E. Boehm-Courjault, M. Cantoni, S. Mischler, A. Beaudouin, W. Chitty, J.-P. Vernot, EBSD, SEM and FIB characterisation of subsurface deformation during tribocorrosion of stainless steel in sulphuric acid. Wear **269**, 383–393 (2010)
3. A.H. Zavieh, N. Espallargas, The effect of friction modifiers on tribocorrosion and tribocorrosion-fatigue of austenitic stainless steel. Tribol. Int. **111**, 138–147 (2017)
4. R. Büscher, B. Gleising, W. Dudzinski, A. Fischer, The effects of subsurface deformation on the sliding wear behaviour of a microtextured high-nitrogen steel surface. Wear **257**, 284–291 (2004)

5. A.H. Zavieh, N. Espallargas, Effect of 4-point bending and normal load on the tribocorrosion-fatigue (multi-degradation) of stainless steels. Tribol. Int. **99**, 96–106 (2016)
6. S. Guadalupe Maldonado, S. Mischler, M. Cantoni, W.J. Chitty, D. Hertz, Mechanical and chemical mechanisms in the tribocorrosion of a stellite type alloy. Wear **308**, 213–221 (2013)
7. N. Espallargas, S. Mischler, Tribocorrosion behaviour of overlay welded Ni–Cr 625 alloy in sulphuric and nitric acids: electrochemical and chemical effects. Tribol. Int. **43**, 1209–1217 (2010)
8. D. Landolt, *Corrosion and Surface Chemistry of Metals* (EPFL Press, Lausanne, 2007)
9. J. Perret, Modélisation de la tribocorrosion d'aciers inoxydables dans l'eau à haute pression et haute température (2010)

Chapter 8
Case Studies

8.1 Tribocorrosion in Pressurized Water Reactors (PWR)

Pressurized water reactors (PWR) are the main constituents of the majority of nuclear power plants worldwide. The design of PWR includes a number of mechanical components that are exposed not only to radiation but also to the water (used as moderator and coolant) kept typically at 300 °C and at pressures in the range of 150 bar. Under these conditions most of the metals used in PWR undergo corrosion resulting in the build-up of metal oxide films on the metal surface. Thanks to their low solubility, these oxide films are several micrometres thick (much thicker than the passive films found in aqueous solutions at room temperature, which are few nanometres). Some components are also subject to loading and relative movements resulting in sliding and impact wear. The combination of corrosion by water and these mechanical loadings produce a tribocorrosion situation.

One example of those components is the rod cluster control assembly (RCCA). The function of RCCA consists of intercalating between fuel rods and stainless-steel rods filled with neutron absorbing material. By controlling the relative height between fuel and control rods, it is therefore possible to adjust the supplied power of the reactor to the desired level. In RCCA, the rods, typically made out of stainless steel, are fastened to the spider and guided to the core of the reactor. The RCCA is positioned by the control rod drive mechanism (CRDM) through gripper latch arms (Fig. 8.1) producing a step-by-step sliding movement which frequency and amplitude depend on the need of regulating the reactor power and thus subject to the demand in electrical power over the day. To minimize latch arm degradation, the contacting parts of these mechanisms were made out of Stellite-type alloys (CoCrMo alloys) known for their resistance to wear and seizure.

Nevertheless, unexpected high wear occurred in reactors with respect to the accelerated tests carried out in full-sized loop tests for materials selection purposes [1]. The polished appearance of the worn surfaces suggested that tribochemical effects were responsible for this abnormal wear more than the mechanical-driven phenomena such as abrasion or impact wear. A characteristic feature of tribochemical effects is

© The Author(s), under exclusive license to Springer Nature Switzerland AG 2020
A. Igual Munoz et al., *Tribocorrosion*, SpringerBriefs in Applied
Sciences and Technology, https://doi.org/10.1007/978-3-030-48107-0_8

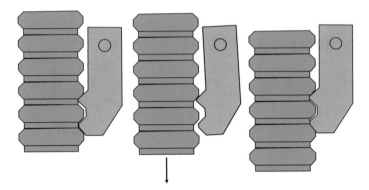

Fig. 8.1 Schematic view of the step-like movement of the latch arm in control rod drive mechanism (CRDM) of nuclear reactor. The whole system operates in pressurized high temperature water

that they are, unlikely mechanical effects, time-dependent. Indeed, it was observed that the severity of wear (wear depth per movement) correlated well with the latency time between step movements of the CRDM, as shown in Fig. 8.2. This dependency is easily explainable if one considers that wear proceeds mainly by the detachment of the brittle oxide formed by corrosion on the Stellite surface. Longer latency time allows more time for corrosion to occur and therefore thicker oxide layer forms on the metal surface. This means in turn that at each step more oxide is removed, leading thus to a larger wear. The loop tests were run in identical chemical and mechanical conditions as nuclear reactor but at higher frequency in order to obtain faster results. However, the small latency time limited the oxide growth and thus led to smaller wear rates than what was found in nuclear power plants under real operational conditions.

Fig. 8.2 Wear rate versus mean time between steps

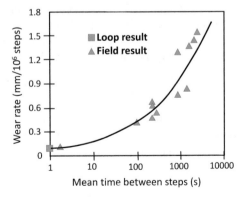

8.2   Tribocorrosion and Chemo-Mechanical Polishing of Tungsten

Tribocorrosion is usually considered as a negative phenomenon leading to the degradation of materials. However, tribocorrosion can be advantageously exploited in engineering practice for device manufacturing and surface finishing. The chemical–mechanical polishing (CMP) process widely used in the fabrication of integrated circuits (IC) constitutes an example of a technological exploitation of tribocorrosion.

The purpose of CMP in the fabrication of ICs is to planarize, by removing excess deposited material, the multilayer interconnect structures (including metallic and insulator layers) grown on silicon wafers. The process involves sliding the functionalized surface of the wafer against a softer pad (polymer, textile) while pouring a slurry composed of suspended abrasive nanoparticles, a dissolved oxidizing agent and other additives (Fig. 8.3). The abrasive nanoparticles trapped between the pad and the wafer are supposed to gently abrade the asperities of the interconnect structure and remove the passive film present on the metal. The depassivated area will corrode till the passive film forms again on reaction with the oxidizing agent. Repeated film removal, corrosion and repassivation cycle lead to surface polishing by the detachment of material from asperities through a tribocorrosion process which in metal CMP is known as Kaufmann's mechanism [2].

As any tribocorrosion phenomenon, the CMP material removal rate (RR, unit thickness per unit time) depends on the interaction of a number of mechanical, material and chemical factors. Not surprisingly, due to its complexity, optimization of the CMP process has been carried out by empirically adjusting pressure, velocity, abrasive particle shape and concentration and chemical composition of the slurry. Mechanical factors are taken into account by the Preston's equation [3] that describe the removal rate as a function of the pressure (P) applied between the wafer and the pad and the sliding velocity (v_s) established in the contact (Eq. 8.1).

Fig. 8.3 Schematic view of the polishing set-up used for CMP

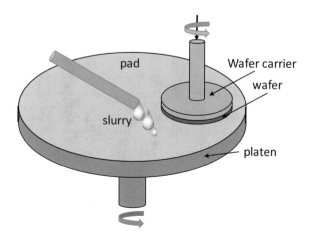

$$RR = kPv_s \qquad (8.1)$$

The Preston's equation is a simple and reasonably well accurate model. However, it does not consider the role of chemistry. Indeed, as shown in Fig. 8.4 for the case of tungsten, the slurry chemistry and, in particular, its oxidizing power play a crucial role. Figure 8.4 shows the RR of tungsten deposited on a wafer and subject to CMP in a series of slurries containing the same abrasive particle but different pH, oxidizing agents (KIO_3 or H_2O_2), concentration of oxidizing agent or oxidation catalysers ($FeNO_3$). For comparison the removal rate obtained using the commercial slurry (CS) of undisclosed composition is given. Figure 8.4 clearly shows the variability of removal rates depending on slurry chemistry. This is not surprising since the CMP of tungsten relies on wear-accelerated corrosion as described before.

In an attempt to rationalize the data shown in Fig. 8.4, Stojadinovic et al. [4] assumed that the cyclic passive film removal and rebuilding were the determining mechanisms controlling material removal during CMP of tungsten (Kaufamn's mechanism). Accordingly, the removal rate, that is the wear-accelerated corrosion rate, should be proportional to the passivation charge density as described by Eq. 4.1 in Chap. 4. In order to verify this, those authors measured the passivation charge density Q_p using independent potential step measurements directly in the studied slurries. The measurement of Q_p consisted of switching suddenly the potential applied to a tungsten electrode from a cathodic (passive free potential) to the arbitrarily selected passive potential of 0.85 V_{SHE} (corresponding roughly to the open circuit potential measured in the tested slurries). The current (the metal oxidation rate) was simultaneously measured by the potentiostat. Q_p was obtained by integrating the current

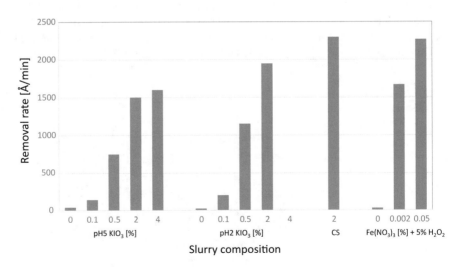

Fig. 8.4 Removal rate during CMP of tungsten in slurries of identical abrasive particles but with different oxidizing agents. CS is a reference slurry containing the same abrasive but undisclosed oxidizing agent. Data from [4]

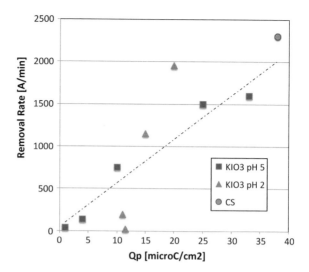

Fig. 8.5 Rationalization of CMP removal rates from Fig. 8.4 through the passivation charge density Q_p. Data from [4]

density over a period from 0 to 0.2 s. Note that in this way it was impossible to determine reliable Q_p values in the H_2O_2 containing solutions. Indeed, in these solutions the reduction of the instable H_2O_2 participates together with metal oxidation to the overall current and therefore the integration of the current values would yield incorrect, underestimated Q_p values.

The measured removal rates plotted as a function of Q_p in Fig. 8.5 reveal that a reasonable correlation exists between these two factors. This confirms that wear-accelerated corrosion is the main material removal mechanisms in agreement with Kaufmann's model. The deviations from a perfect linear trend of Fig. 8.5 indicate that slurry chemistry influences the removal rate also by other mechanisms, for example by modifying the pH (isoelectric point) and/or the surface chemistry (repulsion forces) of the abrasive particles and thus changing their interaction force with the tungsten. Nevertheless, the rationalization of CMP removal rates through tribo-corrosion concepts is a simple and powerful method for screening and tailoring the chemical properties of slurries destined to the simultaneous CMP of different metals and materials. Moreover, using this tribocorrosion approach the Preston's equation could be extended to include the effect of the oxidizing power on the removal rate [4].

8.3 Tribocorrosion in Oil/Water Lubricant Emulsions

Aluminium cans used as beverage containers are formed through drawn and wall ironing (DWI). In this process, cylindrical punch pushes an aluminium preform through a series of dies (Fig. 8.6) in order to form the aluminium in the nearly final can shape (the upper part still needs cutting before applying the cover) and provide the high-quality surface finish required for labelling. Any minimal deviation

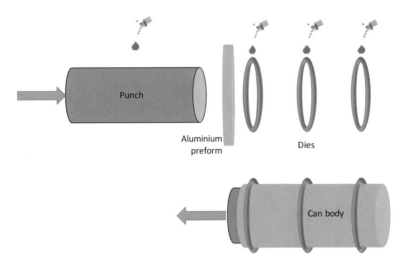

Fig. 8.6 Schematic view of the drawn and wall ironing (DWI) fabrication of beverage cans

of the original geometry of the punch or of the dies results in inhomogeneous wall thickness and other unacceptable surface defects of the produced cans. It is therefore of outermost importance to avoid surface damage of the DWI tools that could occur as a consequence of wear and corrosion.

Corrosion protection is achieved by adding corrosion inhibitors to the oil in water emulsions used as lubricant in DWI. Wear of the tool results from the sliding against the deforming aluminium. Tool materials are hard (carbide alloys such as WC-Co or high-speed steel) and exhibit also high toughness to cope with impact loading. The presence of lubricant and the smoothness of the tool surface prevent adhesion of aluminium on the punch and on the dies. Despite all these measures there is still a need to improve the tool durability in order to reduce the costs of cans fabrication. The can drawing process has been the object of continuous improvement in terms of lubrication, mechanical properties of materials and corrosion efficiency of the inhibitors. However, little attention was paid to possible tribocorrosion effects.

A first simple approach to assess the potential appearance of tribocorrosion effects is the measurement of polarization curve of the tool steel in the lubricant. As shown in Fig. 8.7, the polarization curve of the steel presents in the anodic domain a clear plateau indicative of a passive behaviour of the steel. The polarization curve can also be measured by a simple solution of using the same pH as the lubricant but without inhibitors. Absence of a passive plateau in this simple solution shows that passivity is induced in the lubricant through passivating inhibitors such as borates. Because passive metals are prone to severe wear-accelerated corrosion, the polarization curve strongly suggests that indeed tribocorrosion could play a significant role in the degradation of the tools.

In order to support this hypothesis, a simple potentiostatic tribocorrosion test (see Chap. 5) was carried out using a laboratory tribometer. The tool steel flat was rubbed

Fig. 8.7 Polarization curve of high-speed steel in a 5% oil in water emulsion used for lubricating drawn and wall ironing tools used for manufacturing of beverage cans (blue curve). The comparison with a solution (green curve) of similar corrosivity (same pH) reveals that the lubricating emulsion contains passivating inhibitors as evidence by the current plateau between -0.05 and 0.765 V_{SHE}

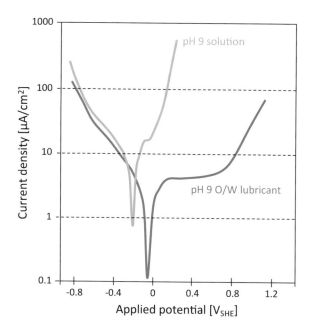

against an alumina ball under a normal load of 5 N in a reciprocating motion of 5 mm amplitude and 5 Hz period. The imposed potential was selected as being 0.965 V_{SHE} more anodic than the open circuit potential measured in the solution before applying the potential. These testing conditions are not supposed to reproduce exactly the drawing conditions but only to assess the potential risk of wear-accelerated corrosion in the system. More details of the experiment are given in [5]. Ideally, the open circuit potential should be applied as can be considered as the one that best represents the real conditions. However, at this open circuit potential, the cathodic current interferes with the response in current during potentiostatic tests and therefore the extent of wear-accelerated corrosion cannot be determined correctly. The anodic polarization of 0.965 V_{SHE} reduces significantly the influence of the cathodic reaction on the measured current [6].

Figure 8.8 shows the evolution of the current during the tribocorrosion potentiostatic test. After the application of potential (time 0) the current shows an exponential drop typical of passive film growth. At the onset of rubbing the current shows a sharp increase indicating activation of wear-accelerated corrosion. After the end of the rubbing the current decreases again to a lower value.

Figure 8.8 clearly shows that the system HSS/oil in water emulsion is sensitive to wear-accelerated corrosion. In order to quantify the extent of wear-accelerated corrosion, one can calculate the current density at the end of rubbing by dividing the current drop observed at the end of rubbing by the area of the wear track. This value needs to be compared to the current density in the absence of rubbing that can be obtained by dividing the current value after the end of rubbing by the area of the sample exposed to the solution. In the present case the wear-accelerated corrosion

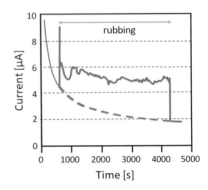

Fig. 8.8 Evolution of the current during the potentiostatic tribocorrosion experiment with an alumina ball on HSS steel contact immersed in a 5% oil in water emulsion under applied potential of 0.965 V_{SHE}. Dotted line: evolution of current in the absence of rubbing by interpolation of the values measured before and after rubbing

reaches values of approximately 0.1 mA/cm^2 while the corrosion rate is three orders of magnitude lower. This increase in corrosion is due to the periodic abrasion and rebuilding of the passive film formed by the reaction between the corrosion inhibitor and the metal.

This simple experiment has shown that in principle the passivating corrosion inhibitors are harmful for this specific tribocorrosion application. The quantitative extrapolation of these tribometer results of the industrial configuration is difficult and would require a more consistent experimental campaign. Nevertheless, mitigation strategies can be already obtained from this simple experiment. For example, one possibility would be to move from passivating to adsorption inhibitors, that is organic molecular, that adsorb on the metal surface without reacting chemically with it. This avoids the removal of oxidized metal during rubbing as in the case of passivating inhibitors. Another option would consist in using more corrosion-resistant materials not requiring inhibitors.

8.4   Hip Joints Simulator Electrochemically Instrumented and Diagram for Open Circuit Potential (OCP) Use

Arthroplasty is a common orthopaedic surgical practice carried out to relieve pain and/or recover the functionality of body joints affected by arthritis or some other type of trauma. It consists of replacing the natural joint by an artificial one and reconstructing the requirements of the original one. The most typical substitutional prosthesis is composed of a stem which is inserted in the femur, a ball attached to the top of the stem and an acetabulum which is anchored into the pelvis and the ball. Metals (i.e. CoCrMo alloys, titanium alloys and stainless steels) are broadly used materials for designing the artificial joint replacement because of their mechanical properties and biocompatibility. Although they are selected because they fulfil the above-mentioned criteria, 85% of artificial knees, for example, have been reported in the last 10–20 years. This constitutes a very important clinical issue for young and active patients who need artificial joints. The analysis of retrievals (Fig. 8.9

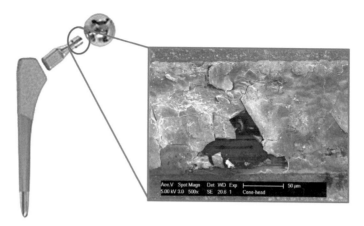

Fig. 8.9 Hip joint prostheses (left) and detail of a retrieved titanium cone coupled to a CoCrMo head where a cracked surface film containing Cr and Mo oxide was observed

shows an example of a titanium cone of a hip joint) has demonstrated that one of the causes of failure is the damage caused by the repeated cyclic loading and motion of the joint in a synovial fluid which is an extremely corrosive environment, thus tribocorrosion has been identified as one of the causes of implant failure. Chemical analysis (EDX) of the worn areas in the explant showed cracked films containing Cr and Mo oxides from the head, which can be only generated as a consequence of the corrosion phenomenon combined to a mechanical action (fretting).

In order to assess this tribocorrosion problem, the behaviour of the materials used for joint replacement can be tested as real components in commercial instruments and simulators, under mechanical conditions similar to those found in the human body, and also using laboratory tribometers (idealized conditions). In both experimental configurations, due to the electrochemical nature of tribocorrosion, together with the material and mechanical conditions, the electrochemical ones (i.e. potential) should be known and/or controlled. As described in Chap. 6, instrumented simulators with the possibility to monitor the open circuit potential during operation are nowadays available after being first developed in Leeds [7]. The registration of the open circuit potential simultaneously to the applied load and the frictional force allowed for monitoring the effect of wear on the corrosion response. The open circuit potential of biomedical alloys such as CoCrMo in a joint simulator represents initially (no operation) the passive state of the CoCrMo. When the simulator starts operating the mechanical loading causes a cyclic removal of the passive film formed on the CoCrMo. Two main zones are then generated in the joint, a worn area in the contact without passive film and an unworn area which remains with the passive film. The potentials of these zones are different. The depassivated or worn zone has lower potential (anode) and the unworn and passive one has higher potential (cathode). Therefore, at the onset of operating, an OCP change (decay) is typically recorded in simulators as a consequence of the galvanic couple formed between the worn and

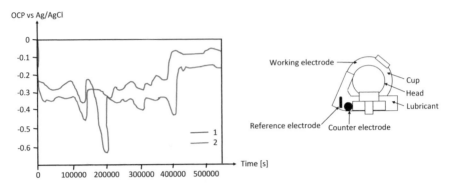

Fig. 8.10 OCP evolution with time (left) of a MoM hip joint tested in a simulator electrochemically instrumented (right). Two different tests are represented. From [7]

unworn areas. And this potential decay is indicative of the mechanical detachment of material.

Figure 8.10 shows an example of the open circuit potential variation during operation of a MoM hip joint simulator in which the head and the cup was made by CoCrMo alloy. The OCP showed an abrupt decrease when the mechanical loading starts, similarly to what was observed in laboratory tribometers (Chap. 6). This potential decay is a consequence of the mechanical damage of the hip joint surface and the cyclic removal of the passive film formed on the CoCrMo alloy generating a galvanic coupling in between worn and unworn areas of the joint.

In order to relate the potential decay registered in Fig. 8.10 with a material loss (metal lost by wear-accelerated corrosion), theoretical tools such as the model developed by Vieira et al. [8] can be used (see Chap. 4). In their model, the OCP evolution with time was expressed as a function of the anodic current rate (i_a, metal ion release) in the wear track caused by the mechanical detachment of material, thus by the detachment of the passive film in the contact zone and the acceleration of active corrosion by wear. Equation 4.2 (Chap. 4) can be represented in the form of tribocorrosion diagrams as the one shown in Fig. 8.11. These diagrams, which are material and environment specific, graphically represent the logarithm of the anodic current density coming from the worn parts of the material (i_a) as a function of the measured electrode potential, E_c, for different values of the worn and unworn area ratio (A_a/A_c ratio, respectively).

The diagram can be used to estimate i_a from a known contact geometry (A_a is the worn surface area, A_c is the overall sample surface area) and E_c (potential measured during rubbing at open circuit). This current density is directly proportional (according to Faraday's law, Chap. 2) to the material loss due to wear-accelerated corrosion. Thus, with very simple experimental measurement (monitoring electrode potential of a metallic surface versus a reference electrode, OCP) wear-accelerated corrosion can be easily quantified. Typical wear rate during the run-in of joints measured in simulators is between 10 and 50 μm/Mcycles and the wear-accelerated

Fig. 8.11 Graphical representation of Eq. 4.2 with electrochemical parameters (CoCrMo alloys, Hank's solution, $E_{corr} = 0.38$ VSHE, $a_c = 2.06$ V and $b_c = 0.35$). Numbers in italic indicate the A_a/A_c ratio. From [9]

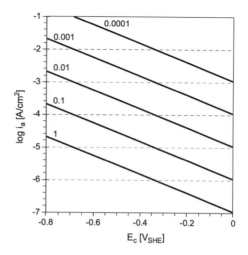

corrosion rate estimated from the measured shift in OCP gives a material loss of 4–40 µm/Mcycles [9]. On the other hand, theoretical wear-accelerated corrosion rate estimated from mechanistic tribocorrosion models is around 25–35 µm/Mcycles. These values allow to obtain several conclusions: (a) wear-accelerated corrosion is one main tribocorrosion mechanism in degradation of joint replacements and (b) OCP monitoring and existing theoretical models allow for quantification and prediction of wear damage.

8.5 The Use of Surface Treatments Against the Tribocorrosion Issues in Biomedical Implants

Metals present a unique combination of mechanical, fatigue and chemical resistance, which make them an efficient choice for implants. Indeed, CoCrMo self-mated artificial joints have been intensively used as biomedical bearing devices. Material loss by corrosion and wear is usually very small in these types of bearing systems. However, the long-term accumulation of metal debris and metal ions in the human body is of great clinical concern [10]. Co- and Cr-ions are considered to be toxic or even carcinogenic and have been shown to promote inflammation and reduced cell activity, which eventually leads to infection and loosening of the implant [11, 12]. In addition to CoCrMo, other metal alloys are used in the human body, such as stainless steels. Therefore, methods to reduce material release from metal alloys implanted in human bodies are sought; in particular, surface treatments have a great technological potential. The surface engineering methods should provide good adhesion, sufficient thickness, microstructural, mechanical and chemical (including biocompatibility) properties suitable for the human body. Assessing the suitability of surface treatments for biomedical applications is best performed, at least in a preliminary face by screening them using tribocorrosion tests in simulated body fluids according to the protocols proposed in earlier chapters.

Figure 8.12 shows the cross-section of titanium surfaces treated by plasma nitriding and vacuum coating (Diamond Like Carbon, DLC) samples [13, 14]. In all cases homogeneous, dense and well-adhered layers were found with the formation of a TiN layer in the plasma nitriding coating.

The tribocorrosion response of the different surface treatments was followed by monitoring the OCP evolution with time of the different samples immersed in a phosphate buffer solution (PBS) sliding against an alumina ball. The potential evolution during rubbing is shown in Fig. 8.13. At the beginning of the tests, the OCP value corresponds to the spontaneous potential established between the sample surface and the electrolyte and is typical of a passive state. When rubbing is initiated, the largest potential drop is observed in the untreated Ti sample, which loses its passive condition as a consequence of the mechanical removal of the passive film. The highest wear was also found in this case. The DLC-coated Ti shows a very small potential change during rubbing and exhibits such low wear that volume loss was

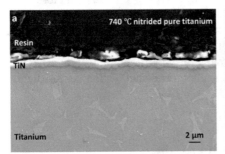
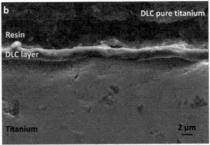

Fig. 8.12 Cross-section SEM images of the **a** 740 C nitrided Ti, **b** DLC Ti

Fig. 8.13 Evolution of the open circuit potential versus time during tribocorrosion tests for the untreated, nitrided- and DLC-Ti sliding against an alumina ball under a normal load of 2 N in a reciprocating motion of 10 mm amplitude and 1 Hz period

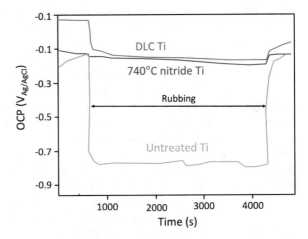

non-detectable under tribocorrosion conditions. The OCP evolution of the plasma nitriding on pure Ti does not vary during the whole tribocorrosion test and produces a thick, dense and homogeneous TiN layer similar to the DLC coating. This provides a comparatively better performance exhibiting excellent electrochemical, tribological and tribocorrosion performance (i.e. very low wear and no galvanic coupling was found).

The simple measurement of the OCP of a surface treatment under tribocorrosion conditions allows for comparing the performance and durability of different coatings. However, timeframe constrains should be carefully considered when choosing surface modifications of alloys for biomedical applications. Long-term testing, especially for coatings such as DLC, should be performed since they might contain porosity that can allow electrolyte penetration leading to damaging adhesion with the substrate and thus catastrophic failures. Of course, for specific biomedical applications, long-term corrosion tests to assess the metal ion release with time would be also of importance before running any in vivo test.

8.6   Tribocorrosion of Hydraulic Cylinders in Offshore Applications

In marine operations heavy machinery is typically involved in critical operations such as mooring, operation of bow ports, material handling by cranes and hoisting systems, riser tensioning, heavy compensation and rotation equipment during drilling. Usually the key components of these systems have slide bearing or sealing surfaces exposed to wear in an offshore chloride-containing environment, such as cylinder piston rods, bearings, chains, hinges, bolts, shafts, swivels, and so on. Typically, these components are routinely replaced after a predetermined number of years in operation (often 5 years) as part of the preventive maintenance (PM) program. These critical tribocomponents can be exposed to the wet marine environment, often also combined with static and dynamic mechanical loading. The reliable operation of these components is of supreme importance not only in order to secure the transport of a multitude of ships carrying cargo and passengers worldwide but also for the exploration and production of natural resources offshore. There are also a large number of naval and civil applications with similar exposure to marine environments such as aircraft carriers, harbours, bridges and storm surge barriers at exposed coastal areas. Emerging renewable energy technologies including offshore wind, tidal and wave energy will meet the same tribocorrosion challenges.

Figure 8.14 shows an example of a marine drilling riser tension system. The vulnerable parts of the system would be the rods and the seals, which are exposed to the marine environment. Sand could get trapped in the seal leading to abrasion of the rod material, while seawater would be splashing the surface of the metallic rod. The metallic rods are typically made of passive alloys such as stainless steel, CoCr or NiCr alloys.

Fig. 8.14 Marine drilling
riser tension system [15]

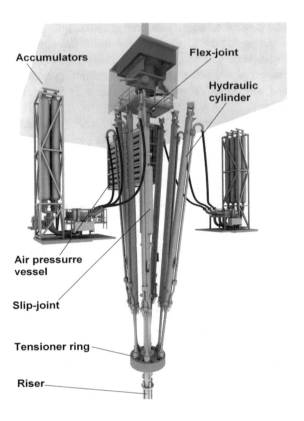

This tribocorrosion case is complex since both fatigue and tribocorrosion are simultaneously acting (i.e. the hydraulic cylinder is subject to cyclic or static bending during operation and major damages are found on mid stroke). Therefore, the first approach is to design a tribometer able to simulate the tribocorrosion and fatigue conditions simultaneously (i.e. simulating both the mechanical and chemical conditions). A laboratory-scale machine for this type of operation was designed and built [16]. A schematic of the machine is shown in Fig. 8.15. The machine is designed with a four-point bending system to apply the cyclic or static bending conditions in a sample that is machined as a bar. The sliding motion is applied in a reciprocating way using alumina ball as counterpart and the normal load is applied by a dead weight. All this is placed inside a testing chamber that is filled with the electrolyte for testing. The electrochemical cell (testing chamber) of the machine is equipped with counter and reference electrodes that are coupled to a potentiostat to monitor in situ and real-time electrochemical parameters such as the OCP and the current of the working electrode (sample to be studied) in a potentiostatic test during the entire test. All the system is properly isolated in order to avoid electrical contacts among the different parts.

The new experimental set-up allows to study simultaneously the influence of the mechanical (i.e. bending fatigue conditions or normal load) and electrochemical

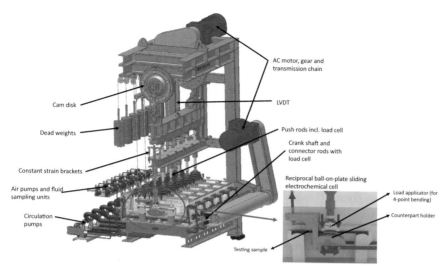

Fig. 8.15 **a** 3D model of the tribocorrosion-fatigue test rig, **b** Electrochemical cell, **c** Wear block with a mounted alumina ball, **d** Four-point bending load applicator [16]

parameters on the material damage. The system allows same testing electrochemical conditions as described in Chap. 6. Figure 8.16 shows the wear volume of an austenitic stainless steel (ASS) tested under applied potential and OCP conditions in seawater. In these tests, the effect of the electrochemical and the fatigue (bending) conditions on the material loss is shown. Ex situ measurement of the tested materials, chemical and microscopical analysis showed differences in the chemical composition and the thickness of the passive film [17].

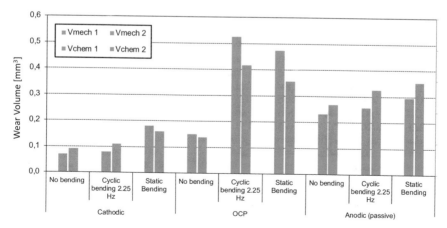

Fig. 8.16 Mechanical and chemical volume loss of ASS at different potentials and bending conditions in 3.4 wt% NaCl and 50 N normal load

Table 8.1 Passive film thickness at OCP and passive potential in all bending conditions inside and outside the wear track for ASS in seawater exposed to fatigue-tribocorrosion at 50 N normal load

	Passive film thickness (nm) versus Ta/Ta$_2$O$_5$ reference inside the wear track [ratio to outside of the wear track]			
	Outside wear track	No bending	Cyclic bending (2.25 Hz)	Static bending
OCP	2.6	7.2 [2.7]	12.1 [4.6]	11.9 [4.6]
Passive	4.1	8.1 [2.0]	10.7 [2.6]	10.1 [2.5]

It is important to note the effect of the potential on the material loss as observed in other triboelectrochemical tests (Chap. 6). However, the results obtained and the careful surface chemical analysis (Table 8.1) of all samples show that fatigue has a great impact on tribocorrosion, especially at OCP. The fatigue action has an impact on the passive film composition and thickness and therefore on the material loss. This careful analysis allows for a design and a choice of the best alloys for the application in order to minimize wear due to the effect of fatigue and tribocorrosion simultaneously. In addition, the electrochemical conditions of the system can be modified in order to control the impact on the degradation. Using the results, an interesting approach was taken where the use of specific additives of the hydraulic fluid in the riser tensioning system was considered in order to modify the passive film composition and thickness [15].

References

1. E. Lemaire, M. Le Calvar, Evidence of tribocorrosion wear in pressurized water reactors. Wear **249**(5–6), 338–344 (2001)
2. F.B. Kaufman, D.B. Thompson, R.E. Brodie, A. Jaso, W.L. Guthrie, D.J. Perason, M.B. Small, J. Electrochem. Soc. **138**, 3460 (1991)
3. P.B. Zantye, A. Kumar, A.K. Sikder, Mater. Sci. Eng. **R45**, 89 (2004)
4. J. Stojadinovic, D. Bouvet, S. Mischler, Prediction of removal rates in chemical–mechanical polishing (CMP) using tribocorrosion modeling. J. Bio Tribo Corros. **2**, 8 (2016)
5. S. Mischler, E. Rosset, D. Landolt, G. Brackenbury, Wear and corrosion in the forming of beverage cans, in *Lubricants and Lubrication*, ed. D. Dowson, C. Taylor, T. Child, G. Dalmaz (Elsevier, Amsterdam, 1995), pp. 119–127
6. D. Landolt, *Corrosion and Surface Chemistry of Metals* (EPFL Press, Lausanne, 2007)
7. J. Hesketh, X. Hu, Y. Yan, D. Dowson, A. Neville, Biotribocorrosion: some electrochemical observations from an instrumented hip joint simulator. Tribol. Int. **59**, 332–338 (2013)
8. A.C. Vieira, L.A. Rocha, N. Papageorgiou, S. Mischler, Mechanical and electrochemical deterioration mechanisms in the tribocorrosion of Al alloys in NaCl and in NaNO3 solutions. Corros. Sci. **54**, 26–35 (2012)
9. S. Mischler, Sliding tribo-corrosion of passive metals: mechanisms and modeling, in *Tribo-Corrosion Research Testing Application*, ed. by P. Blau, J.P. Celis, D. Drees, F. Friedrich (ASTM international, 2013), pp. 1–18
10. A. Igual Munoz, S. Mischler, Effect of the environment on wear ranking and corrosion of biomedical CoCrMo alloys. J. Mater. Sci. Mater. Med. **22**, 437–450 (2011)

11. H. Matusiewicz, Potential release of invivo trace metals from metallic medical implants in the human body: from ions to nanoparticles—a systematic analytical review. Acta Biomater. **10**, 2379–2403 (2014)
12. T.W. Bauer, Particles and periimplant bone resorption. Clin. Orthop. **405**, 138–143 (2002)
13. G.H. Zhao, R.N. Aune, N. Espallargas, Tribocorrosion studies of metallic biomaterials: the effect of plasmanitriding and DLC. J. Mech. Behav. Biomed. Mater. **63**, 100–114 (2016)
14. A. Bazzoni, S. Mischler, N. Espallargas, Tribocorrosion of pulsed plasma-nitrided CoCrMo implant alloy. Tribol. Lett. **49**, 157–167 (2013)
15. A.H. Zavieh, N. Espallargas, The effect of friction modifiers on tribocorrosion and tribocorrosion-fatigue of austenitic stainless steel. Tribol. Int. **111**, 138–147 (2017)
16. C.B. Von der Ohe, R. Johnsen, N. Espallargas, A multi-degradation test rig for studying the synergy effects of tribocorrosion interacting with 4-point static and cyclic bending. Wear **271**, 2978–2990 (2011)
17. H. Zavieh, N. Espallargas, The role of surface chemistry and fatigue on tribocorrosion of austenitic stainless steel. Tribol. Int **103**, 368–78 (2016)

Annex A
Corrosion Rates

Different ways to express corrosion rates can be summarized as follows:

Current density i (A/m^2)

$$i = I/\mathbf{A}_{\text{electrode}}$$

Number of moles n_m which react per unit surface and time (mol/m^2 s)

$$n_{\mathrm{m}} = i/(n\,F)$$

Mass loss m_r per unit surface and time (g/m^2 s^1)

$$m_r = i\,M_r\,/(nF)$$

Thickness of the material reacted d_r per unit time (mm/s^1)

$$d_r = i\,M_r\,/(nF\rho)$$

Table A.1 shows the conversion factors between the different units. To convert the corrosion rate into the units shown in the first arrow, one should multiply the unit of the first column by the corresponding conversion factor.

As an example, in the following table, typical corrosion rate values for pure metals in specific environment are given (Table A.2).

Table A.1 Conversion factors for corrosion rate calculations [1]

	mol/m²s	mol/cm²s	A/cm²	µA/cm²	mg/dm²day	mm/year
mol/m²s	1	10^{-4}	$9.65 \times 10^4 n$	$9.65 \times 10^6 n$	$8.64 \times 10^5 M$	$3.15 \times 10^4 \frac{M}{\rho}$
mol/cm²s	10^4	1	$9.65 \times 10^8 n$	$9.65 \times 10^{10} n$	$8.64 \times 10^9 M$	$3.15 \times 10^8 \frac{M}{\rho}$
A/cm²	$\frac{1.04 \times 10^{-5}}{n}$	$\frac{1.04 \times 10^{-9}}{n}$	1	100	$8.96 \frac{M}{n}$	$0.327 \frac{M}{n\rho}$
µA/cm²	$\frac{1.04 \times 10^{-7}}{n}$	$\frac{1.04 \times 10^{-11}}{n}$	0.01	1	$8.96 \times 10^{-2} \frac{M}{n}$	$3.27 \times 10^{-3} \frac{M}{n\rho}$
mg/dm²day	$\frac{1.16 \times 10^{-6}}{M}$	$\frac{1.16 \times 10^{-10}}{M}$	$0.112 \frac{n}{M}$	$11.2 \frac{n}{M}$	1	$\frac{3.65 \times 10^{-2}}{\rho}$
mm/year	$3.17 \times 10^{-5} \frac{\rho}{M}$	$3.17 \times 10^{-9} \frac{\rho}{M}$	$3.06 \frac{n\rho}{M}$	$306 \frac{n\rho}{M}$	$27.4\, \rho$	1

Table A.2 Typical corrosion rates of metals in different environments

Material	Environment	Corrosion rate ($mg/dm^2/day$)
Low carbon steel	Seawater	16.0
Low alloyed steel 3% Cr	Seawater	31.7
SS 316	10% H_2SO_4	nil
	20% H_2SO_4	50
SS304	10% H_2SO_4	400
	20% H_2SO_4	1000
Al 3004-H14	Distilled water pH 2.7	32
	Distilled water pH 6	1.6
	Distilled water pH 10	12.8
Copper alloy C70600	Seawater	0.64
Titanium grade 12	H_2SO_4 10%	47.0
	Seawater	Nil

Annex B
Synopsis of Electrode Potential

The electrode potential is a very important parameter constituting the driving force for electrochemical reactions. It cannot be obtained directly, and therefore different concepts are associated to this electrode potential depending whether the potential is measured, imposed or calculated.

Open Circuit Potential (OCP):

The open circuit potential (OCP) is measured by connecting through a high impedance voltmeter, with the metal under investigation (working electrode, WE) and a reference electrode (RE) (Fig. B.1). Since at OCP no external current flows, this implies that any anodic current (e.g. metal oxidation) is compensated by a cathodic current (e.g. proton reduction).

Applied potential (E):

The potential is usually applied by connecting a three-electrode cell set-up involving a working electrode (WE, metal under study), reference (RE) and counter electrode (CE) to an electronic device called potentiostat that maintains a selected applied potential between the WE and the RE by passing an appropriate current between the WE and the CE (Fig. B.2).

Corrosion potential (E_{corr}):

The corrosion potential (E_{corr}) corresponds to the potential measured in a polarization curve when the current changes sign, for example from negative to positive values. At this potential the reaction rate (current) of the anodic and the cathodic reactions is the same. The polarization curve is obtained by scanning the applied potential while measuring the response in current. Note that the obtained value of E_{corr} depends very much on the scan rate and scan direction. For actively corroding materials, E_{corr} and OCP are usually the same, and therefore the two terms are considered equivalent and used indistinctively. However, in case of passive metals, E_{corr} and OCP can differ significantly, and therefore in this book the distinction will be kept. Figure B.3 shows an example of a polarization curve.

© The Author(s), under exclusive license to Springer Nature Switzerland AG 2020
A. Igual Munoz et al., *Tribocorrosion*, SpringerBriefs in Applied
Sciences and Technology, https://doi.org/10.1007/978-3-030-48107-0

Fig. B.1 Experimental
set-up for measuring the
OCP

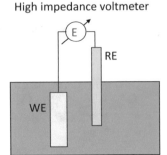

Fig. B.2 Three-electrode
cell set-up

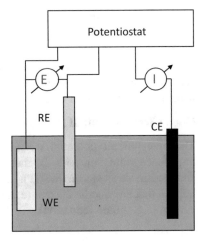

Reversible or equilibrium potential (E_{rev}):

This potential corresponds to the potential at which the intensity (rate) of the oxidation and reduction reactions are equal. In this case an equilibrium state is attained as described by the following reaction:

$$M^{n+} + ne^- \leftrightarrow M \tag{1}$$

This equilibrium potential can be calculated by Nernst equation derived from thermodynamical concepts. Equation B.1 shows the Nernst equation applied to a metal in equilibrium with its ions (reaction 1)

$$E_{rev} = E^0 + RT/nF \ln a_{Mn+}/a_M \tag{B.1}$$

where E^0 is the standard potential, R is the gas constant, T is the temperature, n is the oxidation number, F is the Faraday's constant and a_{Mn+} and a_M are the activities of the metal ions and metal, respectively.

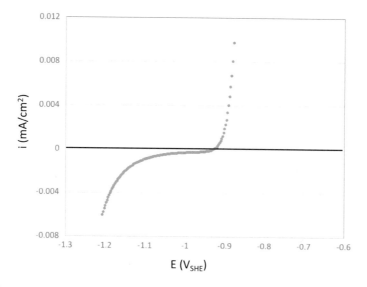

Fig. B.3 Polarization curve E-i of Sn in citrate buffer; pH 4.5

Table B.1 Standard electrode potentials

Electrode reaction	E^0 (V)
$Mg^{2+} + 2e^- \leftrightarrow Mg$	-3.045
$Al^{3+} + 3e^- \leftrightarrow Al$	-1.670
$Ti^{2+} + 2e^- \leftrightarrow Ti$	-1.630
$Cr^{3+} + 3e^- \leftrightarrow Cr$	-0.900
$Zn^{2+} + 2e^- \leftrightarrow Zn$	-0.760
$Fe^{2+} + 2e^- \leftrightarrow Fe$	-0.440
$Ni^{2+} + 2e^- \leftrightarrow Ni$	-0.257
$2H^+ + 2e^- \leftrightarrow H_2$	0.000
$Cu^{2+} + 2e^- \leftrightarrow Cu$	0.340
$Ag^+ + e^- \leftrightarrow Ag$	0.799

Table B.1 shows the standard potentials for electrode reactions at 25 °C, 1 M activity of soluble species and 1 atm. activity of gaseous species.

Table B.2 shows the simplified (25 °C and pressure 1 atm.) Nernst equations for the main corrosion reactions. The equations can be further simplified by assuming the activity equal to the molar concentration for dissolved species.

The graphical representation of the reversible potentials [2] for a given metal as a function of pH is called E-pH or Pourbaix diagrams (Fig. B.4 shows a simplified Pourbaix diagram for nickel). These diagrams are useful for identifying regions where corrosion cannot occur (immunity) or when metals corrode actively or are passive.

Table B.2 Simplified Nernst equations for the main corrosion reactions

Reaction	E_{rev} (V)
$M^{n+} + ne^- \leftrightarrow M$	$E_{rev,M} = E_M^0 - 0.059n^{-1}\log a_{Mn+}$
$M + n/2H_2O \leftrightarrow MO_{n/2} + nH^+ + ne^-$	$E_{rev,MOn/2} = E_{MOn/2}^0 - 0.059pH$
$2H^+ + 2e^- \leftrightarrow H_2$	$E_{rev,H2} = -0.059pH$
$O_2 + 2H_2O + 4e^- \leftrightarrow 4OH^-$	$E_{rev,O2} = 1.23 - 0.059pH$

Fig. B.4 Simplified
Pourbaix diagram for nickel

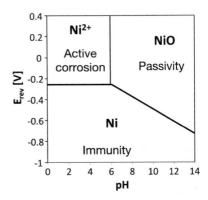

However, these diagrams do not give any information on the corrosion reaction rates and do not consider metastable states. Furthermore, corrosion is a non-equilibrium (irreversible) process so that thermodynamic approaches such as Pourbaix diagrams are of little or no use for estimating the electrode potential.

Annex C
Reference Electrodes

Definition of typical reference electrodes is listed in Table C.1 together with their corresponding electrolyte and potential versus the standard hydrogen electrode (SHE). By convention, the value of zero was assigned to the "hydrogen electrode" which designates the following electrode reaction $2H^+ + 2e^- = H_2$ under standard conditions ($P_{H2} = 1$ bar, $T = 298$ K, $a_{H+} = 1$).

Table C.1 Main reference electrodes and the corresponding potential values with respect to the SHE

Electrode	Electrolyte	Potential (V)
Calomel	Saturated KCl	0.241
	1 M KCl	0.281
	0.1MKCl	0.333
Mercury sulphate	Saturated K_2SO_4	0.658
Mercury oxide	1M NaOH	0.098
Silver/silver chloride	Saturated KCl	0.195
Copper sulphate	Saturated $CuSO_4$	0.316